D1276066

SCIENCE *in* UTOPIA

A MIGHTY DESIGN

SCIENCE
in
UTOPIA

A MIGHTY DESIGN

by Nell Eurich

HARVARD UNIVERSITY PRESS

Cambridge, Massachusetts 1967

Distributed in Great Britain by Oxford University Press, London
Library of Congress Catalog Card Number 67-14339
Printed in the United States of America

To Julie and Don

PREFACE

MANY STUDIES have been made of utopias—man's dreams of a better world—from social, economic, and political points of view, but none exists that concentrates on the role of science in providing the anticipated blessings. Yet our actual progress over the last 300 years has depended heavily on scientific advancement. For this reason I have examined how, why, and when writers of utopias placed their hope in scientific discoveries and their application.

The attempt at synthesis and continuity of presentation, which covers several centuries, may be of value in an age of "exploding knowledge," though it cannot do justice to each period or field of study. In the notes I have indicated for the interested reader where one may find more detailed information on particular subjects.

Limits have been set in the following ways: within the broad spectrum of contributory elements to scientific growth, only the most important components have been singled out for attention. And similarly, on the utopian side, focus is on this one aspect and other facets of utopians' lives only as they relate to it. I have omitted such things as communistic leanings in utopian lands where common ownership of goods and a planned economy are frequent. After Plato's *Republic* there was nothing new in the idea of community of property. In the history of utopian literature this has only recently become a politically charged issue; hence I am not concerned with Marxists' repudiations or interpretations of utopias.

To avoid distortion, the author's own words are repeated as often as possible and placed in perspective both with regard to his other writings and the long view of man's search for the means to a better life. His ideas may reflect many others before him, but he may still be the first to express them in utopian literature. In this sense the study reflects the history of ideas as seen in this literary genre.

For the idea of pursuing this subject, I am especially indebted to the teaching and writing of Professor Marjorie Hope Nicolson, who criticized an early draft of the study several years ago. Dr. J. Max Patrick was most helpful when I began the study of utopias; Dr. Moses Hadas and Dr. René Dubos also pointed out mistakes and difficulties, and offered wise suggestions.

A research grant from New York University enabled me to have the assistance of Dr. John F. C. Richards in checking the translation of Andreae's work from the Latin and Dr. Renata von Stoephasius in the translation of German works concerning Andreae. In addition, Dr. Daniel J. Donno of City College has made for this study a new translation, the first in English, from the earliest manuscript of Campanella's *Città del Sole*. His translation in full is to be separately published. To Alvin C. Eurich, my husband, I owe that peculiar type of psychological support so necessary to one immersed in a utopian world.

Nell Eurich

VASSAR COLLEGE
MARCH 1967

CONTENTS

SCIENCE *in* UTOPIA

A MIGHTY DESIGN

INTRODUCTION

Man has always had his Isle of the Blest, his Elysian Fields, Hesperides, or Garden of Eden, the Golden Age of the past or the Millennium to come in the future. Perhaps we cannot live without such dreams. In any case, we seldom tire of reading a projection into the future or imagining a better world that seems to solve our present problems. This type of sublimation has been a part of the human race from earliest times and has appeared in the literature of all peoples, whether indigenously or by contact and influence. In folklore and the songs of childhood, the romantic theme of an ideal state is widespread and a constant favorite. It is found in the very ancient Sumerian legend—known as early as 3000 B.C.—the epic of Gilgamesh and his search for the mystic isle where the old sage Utnapishtim dwelt forever. And it is pictured today in the popular song, "On the Good Ship Lollipop," which takes the child to Peppermint Bay where bonbons play. Nothing is impossible in this magic land; whether in child or man, desires are satisfied and dreams fulfilled.

Perfect dream worlds have drawn the critics' fire with charges of being escapist, unrealistic "utopias," impossible states of fantasy peopled by nonexistent human beings. This was the reaction Shakespeare gave to those who heard Gonzalo's plan to govern his utopia in such perfection as "t' excel the golden age." There was to be no need for traffic, magistrates, letters, or even occupations:

> All things in common nature should produce
> Without sweat or endeavour: treason, felony,
> Sword, pike, knife, gun, or need of any engine,

Would I not have; but nature should bring forth
Of its own kind, all foison, all abundance,
To feed my innocent people.[1]

And the three cynical, worldly gentlemen who listened to Gonzalo's romantic description decided that the paradise of plenty was nonsense. Antonio commented: " 'T was you we laughed at," and Alonso said: "Prithee, no more; thou dost talk nothing to me." They thought Gonzalo's dream of the ideal commonwealth a most amusing trifle. "Utopia," the word coined by Sir Thomas More in the early 16th century, meant "no place,"[2] and gradually became synonymous with Never-Never Land for many writers and readers.

Long before More introduced the word or Shakespeare's actors sneered at Gonzalo, however, the same negative opinion was registered against idealistic writers who projected new and improved states for man to live in happily ever after. Even Gilgamesh, who achieved great fame as a hero, was parodied in jokes and forgeries. A fraudulent letter, supposed to have been written by the hero to a king, survives in four copies from 8th-century B.C. tablets. It commands the recipient to send ridiculously large quantities of livestock and metals, together with gold and precious stones weighing at least 30 pounds to make an amulet for Gilgamesh's beloved friend, Enkidu.[3] Similarly, Plato's *Republic* received its full share of attention from those who considered it impractical or an elusive, shimmering dream. Whether Aristophanes deliberately satirized parts of the *Republic* is not certain, but the fact remains that he selected some of the same points for caricature in his *Ecclesiazusae*. In this play the women under Praxagora's leadership cleverly gain control of the government and immediately institute sweeping reforms: community of goods and women. And the outcome is very different indeed from Plato's expectation that these changes would protect the state's interests. Instead, the purpose in the ladies' state is pleasure and enjoyment for all, and general promiscuity results. Further, when the time comes for citizens to march forward in procession and turn over their worldly possessions to the state and thence proceed to the communal dining hall, which is to be very gay with wine, women, and song, oddly enough some hang back and figure out ways to avoid giving away their property! It is clearly a burlesque of the same points Plato proposed, regardless of whether Aristophanes actually heard sections (Books 2–5 incl.) of the *Republic* before or during the writing of his play, which was exhibited first in 392 B.C.[4]

Through the years Plato has continued to be so judged by his critics: for example, in the 2nd century A.D. when Lucian wrote his most amusing *True History*, which satirized all idealistic lore including Homer and the giant figures of classical Greece, he observed that Plato was nowhere to be seen at the "delightful party permanently in progress" in the Elysian Fields, and reported: "I was told later that he had gone to live in his own *Republic*, where he was cheerfully submitting to his own *Laws*."[5] Again centuries later we hear adverse comment on Plato's society and idealistic states in general, this time from the French political philosopher Jean Bodin, when he wrote the *Six livres de la République* (1576). He explained his practical, serious purpose in contrast to the imaginary utopia:

> Yet is it not our intent or purpose to figure out the onely imaginary forme and Idea of a Commonweale, without effect, or substance, as have *Plato*, and Sir *Thomas More* Chauncelor of England, vainely imagined: but so neere as we possibly can precisely to follow the best lawes and rules of the most flourishing cities and Commonweals.[6]

At least on one occasion, John Milton seemed to join in this negative attitude toward utopias when he said in the *Areopagitica* that he had no intention of indulging in speculative schemes of perfection. "To sequester out of the world into Atlantick and Eutopian polities, which never can be drawn into use, will not mend our condition." And in *An Apology* he roughly condemned Joseph Hall's utopian *Mundus Alter et Idem* as "a meer tankard drollery, a venereous parjetory for a stewes." This time, however, Milton also expressed the other

—affirmative—attitude of his century. After castigating Hall's work as contemptible, he paid high tribute in contrast to the constructive, idealistic utopia:

> that grave and noble invention which the greatest and sublimest wits in sundry ages, *Plato in Critias,* and our two famous countreymen, the one in his *Utopia,* the other in his *new Atlantis* chose, I may not say as a feild, but as a mighty Continent wherein to display the largenesse of their spirits by teaching this our world better and exacter things, then were yet known, or us'd.[7]

This "mighty Continent" in which men may express their idealism is our concern in the present study. To be sure, there has always been and will be the cynical point of view perhaps especially goaded by the plethora of superlatives used in describing perfect lands. We frequently encounter the disillusioned sophisticate who laughs derisively at the dreamer's scheme, yet we do not willingly let the ideal die. The enthusiasm and spirit of belief which Shakespeare gave to Miranda in that other type of dream in *The Tempest* (V.i.181–184) is ever renewed and ever possible:

> O, Wonder!
> How many goodly creatures are there here!
> How beauteous mankind is! O brave new world,
> That has such people in 't!

The words do not grow old though it is true that hers is the dream of youth and unspoiled innocence to which her father quietly replies: " 'T is new to thee." With a touch of sadness age may contemplate the faith of youth in a new and beautiful world, but somehow we are careful never to deny it. As Mark Van Doren remarks: "The carcass of the world that age hands on to youth is suddenly not a carcass but a brave new goodly thing."[8] This is the hope of men that survives youth's unlimited faith and continues to express itself eternally in the creation of utopias. It is the same belief that John Donne reaffirmed when considering the illness and mortality of this our present world. Only through the practice of Virtue, the Idea which informed matter for Plato and Donne, could this decaying universe find its rebirth. It is virtue

> Which, from the carcasse of the old world, free,
> Creates a new world, and new creatures bee
> Produc'd: the matter and the stuffe of this,
> Her vertue, and the forme our practice is.[9]

From this realm of idealistic thought utopias emerge, positive evidence of man's "largenesse" of spirit and his desire to teach "our world better and exacter things." Here are the visions of man that reshape and remold reality as he would like to see it. Utopias then are fictional, imaginative stories of ideal people living in better societies that exist only in the writer's mind, at least at the time of recording. The author may describe many aspects of the perfect life in which case his work is what we shall call a full-scale utopia; but if he chooses to emphasize one or another element only, if he covers fewer aspects of the society, he writes a partial utopia, half-sister to the complete description. To clarify further, we use the word "utopist" to mean the author, the person writing about the perfect place, while we reserve the term "utopian" for those people inhabiting his dream world or for general use as an adjective.

Most idealists who have composed utopias have adopted certain techniques of presentation that become distinguishing characteristics of the literary genre; for example, authenticity is generally fabricated to give the tone of reality or actual existence to the ideal city. To do this the author may have one reliable informant who saw it all with his own eyes, or he may use extensive details of dimensions and numerical figures to convince the reader. Similarly, he will usually place his utopia so that it is inaccessible to the traveler, and he may borrow from his predecessors an imaginary or geographic location like Atlantis, Ethiopia, Terra Incognita, or the equator. Regardless of his technique, he must create an atmosphere of a living society albeit perfect. Thus an ideal political constitution, while obviously idealistic and including laws for many areas of life, is not a utopia because people do not seem to move and breathe in it. For the same reason an ideal philosophic system fails to become a utopia. Too often the philosopher and political scientist forget to build houses or provide food without which their ideas cannot be absorbed. The utopist projects his ideal in action, describes the daily life and habits of his imaginary inhabitants; and we as readers have the rare opportunity of vicariously experiencing life in the ideal land. Still of course there is considerable latitude in the forms that utopias take, just as there is in any literary genre, which attests to its vitality: it changes with the originality and contribution of various authors. Furthermore, the utopia's origin in the idealistic tradition of letters, which we shall briefly trace, certainly makes its definition no easier. We can only conclude that utopias are imaginary stories of better people and more perfect worlds.

Now the question remains as to their value. When conceived from Miranda's constructive viewpoint, utopias have often expressed goals for society; they have been more than the land of Cockaigne or Gonzalo's escapist vision of a beguiling place. As expressions of the ideal they have served as a flash of perfection ahead, enticing man to strive for it, to improve his way of life. In this sense the ideal state becomes a potential guide for the energy creating civilization and so is directly related to the history of ideals and ideas. The development of thought patterns, if not always historical facts, is reflected in the attitudes held by the author, in the point of departure for his utopian land, and in the details of the society he proposes. Thus the serious utopist produces a commentary on the historical period in which he lives as well as a prophetic projection into the future, which may or may not be realized.

Some writers have seen utopists in a more powerful position as determinants of history. Arthur Morgan in his book *Nowhere Was Somewhere* suggests that "creative vision . . . may become an active cause of events, without which they would not occur. At their best utopias have that gift." And Joyce Hertzler in *The History of Utopian Thought* says utopias have played a great role in shaping human history: "After all Utopia is not a social state, it is a state of mind." Lewis Mumford too has analyzed *The Story of Utopias*, separating the escapist dream from the type of utopia that reconstructs the environment and recommends new habits and values. Although Mumford would start his new state with an organized force of scientists and social scientists on a regional basis, examining local problems, resources, and habits of thought that are vital to the success of a project and often overlooked by the visionary, he endorses the power of the ideal and quotes Anatole France:

> Without the Utopians of other times men would still live in caves, miserable and naked. It was Utopians who traced the lines of the first city. . . . Out of generous dreams come beneficial realities. Utopia is the principle of all progress, and the essay into a better future.[10]

It is not necessary to defend the ideal as a force in history against other determinants such as economic factors, the chance element, or the inevitable cycle of rise and fall; nevertheless, it is important to recognize that without man's efforts toward a better life, without his desire to leave a better world for his children, without his assumption of responsibility in this direction, he becomes a meaningless blob of mere existence. Perhaps then the *process* of building utopias is the main point, not their realization. Oliver Wendell Holmes adds along this line: "The great thing in this world is not so much where we stand as in what direction we are moving." This is the attitude behind Arnold Toynbee's belief that as long as the sights are ahead for a society, as long as it meets the challenge of problems and so develops, it is a growing civilization; but once its view turns to the past, its mimesis patterned on the days gone by, the society becomes conformist and starts on the downhill road. Then another fresh, energetic group facing its problems arises to create for itself a dominant place during its period of growth.

Utopias then are models for mimesis usually timed for the future, and the best of them may well have influenced many readers. We do not claim them to be formative to civilization except as they are related to ideals affecting man's behavior and giving him new aims. Instead, it is our intent simply to trace the course

—the evolution—of some of man's most vital beliefs recorded in utopian literature. The final desire remains constant: happiness and a better life. But the components of that happy life and the means or method to achieve it vary greatly and provide a striking comment on the development of civilization. There were times when virtue was the sole means for reaching a blessed state either here on this earth or in the hereafter; other periods and authors have held that political reorganization was the primary way to create a better world. Still others placed their hopes in economic and social reform or education as the major road leading to the perfect society. And finally science in the modern sense entered to provide the blessings for man in utopia.

To see this development in historical perspective necessitates the survey of early utopian societies: their foundations in idealism, the methods recommended for achieving the ideal, and the answers given for man's problems. From this review the concept of progress emerges, gains adherents, and encourages reforming idealists to project the goal for progress. Within this framework attention will focus in detail on those utopists in the 17th century who first saw the promise of modern science and made it the great hope and means by which utopia was to become reality. As we live in the world they visualized nearly 300 years ago, we may smile at their naïveté but we must respect their vision, still wondering perhaps why the world does not seem to us quite as perfect as they thought it would be.

When they recorded their dreams for scientific progress, the study of science was of course barely beginning. Before the 17th century, the word "science," deriving from the Latin, had always meant knowledge in its broadest sense. What we now call science was then often designated by the general term "natural philosophy," which embraced astronomy, physics, mathematics, chemistry, and mechanics, as well as natural history, including the study of all natural objects. The conquest of nature was the goal, and any study or experiment contributing to that conquest was a legitimate part of "science." By the time the first utopias with strong scientific emphasis appeared, Gilbert, Kepler, Galileo, and Harvey had announced their discoveries. These facts added dramatically to the new thought emerging, the new philosophy about man and the universe that was forced into being by scientific development. A thinking man could no longer accept his inherited beliefs without questioning and subsequent revision. This sort of reaction, proof of the impact of the new thought, is seen in utopia. The utopists' imaginations were immediately challenged by the possibilities of the new science. Their period was not unlike our own as we conquer the universe of outer space, experience the excitement of cosmonauts, send satellites into orbit, and wonder at the new powers to control more than this old, familiar globe. Our sights too require adjustment: the "new learning" of the 20th century must be understood and managed in its many ramifications. Our values too are inevitably affected as we find ourselves even less significant in a universe of many explorable worlds, yet more important as the human mind achieves the unbelievable. Perhaps the thing most to be admired in our 17th-century counterparts, who faced a comparable challenge, is the speed with which they absorbed their "new learning" and creatively projected its benefits for all mankind.

Hence, it seemed to me more than a matter of literary curiosity to examine carefully the utopias in which science initially appeared as the answer to man's problems—but with the background of other utopian solutions which should not be forgotten or lost because of the awesome dominance of science in the present day.

I

THE HEAVENLY PARADISE

"Ah! But a man's reach should
exceed his grasp
Or what's a heaven for?"

—BROWNING

PRIMITIVE PICTURES of man's ideal land, although painted in the words of various languages, resemble one another rather closely and first emerge from a religious context. It seems to matter little whether the writer is pagan or Christian; the descriptive imagery chosen, especially for the early views of a blessed state, remains a literary heritage common to all people. The human mind may simply be limited in the total choice of words when describing the perfect, which is the unknown; so indigenous expressions arise with a more or less stereotyped pattern. Or, as we now know that relationships between ancient civilizations were much closer and more effective in transmitting concepts than was formerly recognized, it may well have resulted from direct contact. Scholarship and archeological discoveries are constantly adding evidence for this theory of interplay between different societies of people in the preclassical world. In either case, the initial visions of a heavenly life appear with surprisingly similar features. Subsequently, of course, the view will be extended, new elements added, and marked differences become apparent, thus changing the entire tone and meaning of life in the imagined world. An understanding of this early development—the original model, its expansion, and the entrance of a dynamic concept—lays the foundation for progress in utopias to come and provides the basis for considering evolution in man's ideals.

The Early Utopian Pattern

Gilgamesh, King of Uruk in Mesopotamia and predecessor of many heroes whose names are better known, was probably first to make the pilgrimage to the land of heart's desire. He traveled nearly 2000 years before Homer's heroes, and his wish was a common one: to live forever, the wish for immortality. On his fantastic journey to find the solution, which might best be called merely wisdom, he walked through terrifying wilderness, killed lions in the mountain passes, and bravely followed the "road of the sun." Curiously, it was the lady of pleasure, the alewife and advocate of merry living, Siduri, who gave him directions for crossing the waters of death, which were the far boundary of the known world. A similar character, Circe, was later chosen by Homer to give instructions to Odysseus for crossing into the Underworld, but Gilgamesh's travels were on the surface of the earth, even though it was an otherworldly surface insofar as geographic places are determinable. With the help of an informed ferryman, Gilgamesh eventually reached Dilmun "in the land *of crossing* . . . the place where the sun rises."[1] Here he found Utnapishtim, the Faraway, who had survived the great deluge and after his trip on the ark was rewarded by the gods with everlasting life. It is further stated in an Assyrian version of the myth that this beautiful city or land (it seems to have been both) was located "at the mouth of the rivers."[2]

The Greeks comparably chose the Springs of Ocean for their Elysian Fields and their Blessed Isles, but they placed them in the west. Their Okeanos, the circling stream, never-ending, and yet signifying the end of the earth was, according to Hesiod, born of Earth in the bed of Heaven, which thus explains its existence at the point where earth joins the realm beyond. Into this stream the heavenly constellations, all except the Bear, regularly bathed in their due course just as the sun was put to rest there. For both Sumerians and Greeks the blessed resided at the mouths or springs of the water where the present world ends. And crossing it led to immortality and a place of blissful rest.

Gilgamesh unfortunately did not describe physical beauty in this ideal land though he expressed surprise that Utnapishtim was lying there indolently, resting upon his back. It is from a tablet at Nippur in Babylonia that we learn the land of Dilmun is pure, clean, and bright. Its waters are abundant and sweet from the "mouth whence issues the water of the earth." The fields are furrowed and give forth grain; there is no sickness, old age, strife, or lament. It is here that even the gods were born from Enki (the Sumerian god of the waters) and Ninhursag (the mother of the land) for

> In Dilmun the raven utters no cries,
> The *ittidu*-bird utters not the cry of the *ittidu*-bird,
> The lion kills not,
> The wolf snatches not the lamb,
> Unknown is the kid-devouring *wild dog*,
> Unknown is the grain-devouring . . . ,
> [*Unknown*] is the . . . widow,
>
> • • •
>
> The dove *droops* not the head,
> The sick-eyed says not "I am sick-eyed,"
> The sick-headed (says) not "I am sick-headed,"
> Its old woman (says) not "I am an old woman,"
> Its old man (says) not "I am an old man."[8]

The inscription further explains that this was true when the world was young and the work of creation had only just begun. Dilmun thus appears to have been an all-inclusive ideal, the original home of the gods, embracing the past age of perfection in the beginning of creation as well as the eternal life to come in the future, the reward for those few chosen by the gods. The Greeks and Hebrews, however, separated the two places by envisioning the Golden Age and Garden of Eden for past perfection and the Elysian Fields or heavenly abode for future happiness.

Regardless of Dilmun's place in the chronology of time or its Greek-like view of a life of ease and its comparable location at the source of the waters, one immediately is reminded of Biblical promises phrased in the same kind of symbolism and sentence structure. In Isaiah 11.1–9 the Messianic kingdom is portrayed when a David reborn shall judge righteously with equity for the poor and meek, and peace shall reign:

> The wolf also shall dwell with the lamb, and the leopard shall lie down with the kid; and the calf and the young lion and the fatling together; and a little child shall lead them. . . . They shall not hurt nor destroy in my holy mountain: for the earth shall be full of the knowledge of the Lord, as the waters cover the sea.

A beautiful simplicity of expression characterizes both descriptions. Although tense shifts from the present in Dilmun to the future tone of Israel's promised kingdom, the very graphic embodiment of the ideals of peace, justice, and equality was practical and certain to gain its effect on the listener or reader. For primitive civilizations, peace among the animals easily represented the absence of strife and destruction in human life. The parallel may, in this instance, result from direct contact, since Sumerians were the first literate people of the area and were conquered by Semitic tribes during the third millennium.

Other types of imagery, of course, were also conceived for early visions. Among literary prophets of the Old Testament, Amos, a shepherd living in the mountaintop village of Tekoa during the mid-8th century B.C., about a generation before Isaiah, had drawn his simple, clear metaphor from his own pastoral life. He too had projected those days of exultation when the Lord will restore the tabernacle of David and

> the plowman shall overtake the reaper, and the treader of grapes him that soweth seed; and the mountains shall drop sweet wine, and the hills shall melt . . . and they shall build the waste cities, and inhabit *them;* and they shall plant vineyards, and drink the wine thereof; they shall make gardens, and eat the fruit of them.[4]

From the more sophisticated Greeks, however, came a flowing statement of the felicity to come; theirs was to be a life of idyllic ease in the perfect climate as first pictured by Homer, who probably lived during or shortly after the time of the Hebrew Amos. In the *Odyssey* (4.563ff) Proteus reassured Menelaus that because he was Jove's son-in-law, the husband of fair Helen, he was singled out for the hero's rewards and would be taken to the blessed plain at the ends of the world.

> There fair-haired Rhadamanthus reigns, and men lead an easier life than anywhere else in the world, for in Elysium there falls not rain, nor hail, nor snow, but Oceanus breathes ever with a West wind that sings softly from the sea, and gives fresh life to all men.

And Homer retained the same view, even repeating his weather conditions, when he described the situation of the gods on Olympus:

> Here no wind beats roughly, and neither rain nor snow can fall; but it abides in everlasting sunshine and in a great peacefulness of light, wherein the blessed gods are illumined for ever and ever.[5]

Thus for Homer at least the vision of immortal life—whether in the heroes' Elysian Fields or the gods' Olympus—was more or less standardized. The description varied little between the two. Moreover, the same general atmosphere pervaded Hesiod's concept of eternity for heroes who

> live untouched by sorrow in the islands of the blessed along the shore of deep swirling Ocean, happy heroes for whom the grain-giving earth bears honey-sweet fruit flourishing thrice a year, far from the deathless gods, and Cronos rules over them.[6]

The heavenly paradise was poetically beautiful and simply described. Whether placed at the mouth of the waters by the Greeks and Sumerians or on Isaiah's holy mountain, it offered eternal life in a refreshing and generous environment where only gentle winds were felt. The harsh, ugly, and rough elements of strife in nature or among men were unknown. Only one significant difference is apparent in the early views of these three peoples: the Sumerians and Greeks reserved their heavens for a few eligible heroes, the semidivine, or gods, while the Hebrews opened their gates to all who survived the great moral cleansing that was to occur before the promised day could be ushered into reality. The hand of justice and vengeance was to precede the Millennium. And although the Sumerians (in the tablet from Nippur) and Hebrews with a close parallel in imagery and style described a more primitive concept in terms of the animal kingdom and agricultural plenty than the Greeks' highly civilized life of pleasant leisure (which Gilgamesh observed in Dilmun), each was meant to signify a state in which man's basic needs were cared for as well as his higher desires for concord and peace. The scene in all was one of fertility and natural abundance. Plenitude symbolized perfection. Each compensated for the unsatisfactory and unpromising conditions of the present. So early portrayals of the heavenly paradise were rather generalized, even stereotyped as in Homer; the literary pattern was brief and descriptively limited; the implications, nevertheless, were immense and extensive, permitting the imagination unlimited opportunity to amplify and add details to the heavenly utopia.

The Expanding Utopian View

Blessed landscapes grew not only in scope and description until the canvas was fairly crowded with lively activities and colorful scenes, but also the heavenly gates were increasingly extended to admit a growing number of fortunate souls. A rainbow of hues began to shine forth in flowers and trees and the dress of the chosen ones as the diffuse, general atmosphere of earlier visions gave way to more specific details. The model that Homer and Hesiod established for the Greeks was greatly enlarged, for example, in Pindar's view of heroes in the Elysian Fields:

> For them the sun shineth in his strength, in the world below, while here 'tis night; and, in meadows red with roses, the space before their city is shaded by the incense-tree, and is laden with golden fruits. . . .
>
> Some of them delight themselves with horses and with wrestling; others with draughts, and with lyres; while beside them bloometh the fair flower of perfect bliss. And o'er that

> lovely land fragrance is ever shed, while they mingle all manner of incense with the far-shining fire on the altars of the gods.[7]

So the blessed now have means to while away eternity in a game of checkers (the British term used in the translation is "draughts"), bodily exercise, or quietly playing the harp. Not only are heroes happily engaged in activities for the first time in this closer view, but the lyrical Pindar (born 520 B.C.) has spread an incense-laden atmosphere over the scene and created flowering meadows and golden fruits. To this Elysian setting Vergil imaginatively added his ideas:

> They came to a land of joy, the green pleasaunces and happy seats of the Blissful Groves. Here an ampler ether clothes the meads with roseate light, and they know their own sun, and stars of their own. Some disport their limbs on the grassy wrestling-ground, vie in sports, and grapple on the yellow sand; some trip it in the dance and chant songs. There, too, the long-robed Orpheus matches their measures with the seven clear notes, striking them now with his fingers, now with his ivory quill.[8]

The blissful state remained one of haunting fragrance and flowers by "crystal rivers flowing placidly to the sea." Such was still Lucian's view in the 2nd century A.D., and although he approached the Island of the Blest in the midst of his huge parody *The True History*, he described beautifully the commonly accepted view.

> As we drew near the island, a marvellous air was wafted to us, exquisitely fragrant, like the scent which Herodotus describes as coming from Arabia Felix. Its sweetness seemed compounded of rose, narcissus, hyacinth, lilies and violets, myrtle and bay and flowering vine. Ravished with the perfume, and hoping for reward of our long toils, we drew slowly near. Then were unfolded to us haven after haven, spacious and sheltered, and crystal rivers flowing placidly to the sea. There were meadows and groves and sweet birds, some singing on the shore, some on the branches; the whole bathed in limpid balmy air. Sweet zephyrs just stirred the woods with their

> breath, and brought whispering melody, delicious, incessant, from the swaying branches. . . . We came upon the guardians of the peace, who bound us with rose-garlands—their strongest fetters—and brought us to the governor. As we went they told us this was the island called of the Blest, and its governor the Cretan Rhadamanthus. . . . There is no night, nor yet bright day; the morning twilight, just before sunrise, gives the best idea of the light that prevails. They have also but one season, perpetual spring, and the wind is always in the west. The country abounds in every kind of flower, in shrubs and garden herbs.[9]

This is the fullest treatment of the Greek vision that has come down to us,[10] and it is not unlike Peter's revelation of the Christian heaven, probably recorded in the same century and obviously modeled on the classical pattern:

> And the Lord showed me a very great country outside of this world, exceeding bright with light, and the air there lighted with the rays of the sun, and the earth itself blooming with unfading flowers and full of spices and plants, fair-flowering and incorruptible and bearing blessed fruit. And so great was the perfume that it was borne thence even unto us. And the dwellers in that place were clad in the raiment of shining angels and their raiment was like unto their country; and angels hovered about them there. And the glory of the dwellers there was equal, and with one voice they sang praises alternately to the Lord God, rejoicing in that place.[11]

Aside from the angels' raiment and God's presence, it sounds very like the older descriptions. Remembering Pindar's heavenly city and meadows of red roses, we need only remove "the incense with the far-shining fire on the altars of the gods" in order to see Peter's apocalyptic vision. And immediately preceding this private view of the heavenly scene vouchsafed to Peter alone, he had with the other disciples seen the astounding beauty of two men standing before the Lord toward the east.

> Their bodies were whiter than any snow and ruddier than any rose; and the red thereof was mingled with the white, and I am utterly unable to express their beauty; for their hair was

> curly and bright and seemly both on their face and shoulders, as it were a wreath woven of spikenard and divers-coloured flowers, or like a rainbow in the sky, such was their seemliness.[12]

These same red and white colors in glorious light and opulence are found earlier in various Greek writings and they appear later as the colors of love in Dante's great visions and in Renaissance gardens and bowers of bliss.

The manuscript of the Petrine Apocalypse was discovered at Akhmîm in Upper Egypt in 1886 and is held to date back to a time between the 8th and 12th centuries though the original, of course, is earlier. In Andrew Rutherford's introduction to *The Revelation of Peter*, he claims it to be the earliest embodiment in Christian literature of the pictorial presentations of heaven and hell, which subsequently have had so widespread and enduring an influence.[13] It is this classical pattern reflected in Peter's vision that provides the basic structure for John Milton's magnificent heaven many centuries later. He embellished the scene with even richer imagery—one arrives at heaven's impressive gates by sailing on a bright sea of jasper or of liquid pearl, wafted by angels—yet the older model persists. Lucian's twilight atmosphere is all-pervasive and music ever present although angelic choirs have replaced Vergil's white-robed Orpheus. Still Plato's singing planets from the Vision of Er remain and join in the musical chorus of Milton's heaven. Certainly the varieties of produce available have not been reduced. When Raphael explains the angels' dietary habits to Adam and Eve, he says that

> in Heav'n the Trees
> Of life ambrosial frutage bear, and vines
> Yield Nectar, though from off the boughs each Morn
> We brush mellifluous Dewes, and find the ground
> Cover'd with pearly grain.[14]

Milton's heaven would have been easily understood and appreciated in the classical world. The ancient audience would have thought it quite proper that after their song and dance, the angels in assigned circles turned to tables piled high

With Angels Food, and rubied Nectar flows:
In Pearl, in Diamond, and massie Gold,
Fruit of delicious Vines, the growth of Heav'n.
On flours repos'd, and with fresh flourets crownd,
They eate, they drink, and in communion sweet
Quaff immortalitie and joy.[15]

Like earlier concepts of heaven, life was simply perfect. The heavenly utopia, located beyond this earthly realm either at its boundary on the shores of Ocean or, as in Peter's and Milton's Christian view, above the firmament, promised bountiful plenty, a life of ease or pleasant activities amid fragrance, flowers, and flourishing fruits. Only additional imagery, description of the setting, and details concerning activities for the chosen ones have come forth in the expanding utopian view. The concept of bliss had not changed. No desire or need was unsatisfied; hence there was no challenge or problem, and progress was unnecessary when one already possessed the goal. Perfection was then static and timeless. This concept may also account for the fact that the literary pattern itself has not developed further. Little has been added to this picture of the heavenly utopia since the 17th century, when the blind poet Milton had his tremendous view of God's kingdom.

A Dynamic Vision

A very different concept, far from the static or timeless attitude reflected in the Greek paradise, was, however, operative beneath the surface of the heaven seen by Peter in the Christian era or the Messianic promise visualized by early Hebrew prophets like Amos and Isaiah. Their descriptive imagery may have resembled the Greek or Sumerian but they were miles away in basic belief. Peter merely adopted the literary picture of Elysian Fields, grafting it onto his fundamentally dynamic view; similarly, Amos portrayed plenty in pastoral symbolism common also in Greek description, and Isaiah described the life of peace in terms of the wolf dwelling with the lamb, comparable to the example from Sumerian literature. But each of the Judaic-Christian visions was projected from a point of view profoundly in contrast to that of their neighbors. The difference becomes clearer in a heaven such as Enoch's, where not even surface description mirrors the classic type. His was a moving, changing series of inspired visions written with a didactic purpose and desire to convince his reader of its truth.

In the *Book of Enoch*, probably compiled early in the reign of Herod or at least during the first two centuries B.C., it is a fiery heaven presented in exhaustive detail and a mood of close study, typical of apocalyptic writers generally.[16] At first glance, it may seem curious that prophetic visionaries should be so painstakingly careful, even to the point of delineating the stones that make up the heavenly pavements, yet it is obvious that concrete itemization contributes to the effect of veracity. Details create the feeling of actual existence, which the prophet wishes to emphasize in order to gain his reader's belief. Thus he draws a verbal blueprint of the kingdom, adding a myriad of visual elements to the literary picture of heavenly utopias. The same motive—to fortify the reader's credulity—undoubtedly caused apocalyptic writers to borrow the authority of highly respected, ancient names like Moses, Abraham, and even Enoch who certainly had slim backing from canonical Scripture. We know only from Genesis 5.21–23 that Enoch "walked with God" and that "God took him," but these two statements were sufficient to make him one of the chosen and hence knowledgeable of God's ways and His promised kingdom.

Thus "Enoch," writing hundreds of years after the time of the Biblical Enoch, who was Noah's great-grandfather, could predict many events with safety, since they had already occurred. For the vision of heaven above, however, this timing was of little value; heaven remains to be seen, so each author must imaginatively project. And Enoch clearly lacked no talent in detailing the mystical vision, seeing many delights for the blessed, and picking up much information in a wide variety of areas. Upon his first entry (he had several visions) into heaven he was overwhelmed by a vibrating flame and reported that he

> drew nigh to a spacious habitation built also with stones of crystal. Its walls too, as well as pavement, were *formed* with stones of crystal, and crystal likewise was the ground. Its roof had the appearance of agitated stars and flashes of lightning; and among them were cherubim of fire in a stormy sky. A flame burned around its walls; and its portal blazed with fire. When I entered into this dwelling, it was hot as fire and cold as ice.[17]

Amid this violent setting the throne of God rose from rivers of flaming fire (in other passages it rests atop a mountain) and no one could look upon His countenance, but Enoch recognized His robe, which was "brighter than the sun and whiter than the snow."

Elsewhere in various views of the heavenly domain he found more peaceful surroundings and examined them with almost microscopic precision. When the heavenly spirits survey all God's works including those on earth, they "behold every tree, how it appears to wither, and every leaf to fall off, except of fourteen trees, which are not deciduous; which wait from the old, to the appearance of the new *leaf*, for two or three winters."[18] Enoch's botanical interest is further evident in the many trees and plants covering the seven mountains in heaven: the cinnamon of sweet odor, papyrus, pure nard (spikenard), and trees "numerous and large" whose "fragrance was agreeable and powerful" and appearance both "varied and elegant." The tree of knowledge, empowered to impart great wisdom and growing in the garden of righteousness, is like a "species of the tamarind tree, bearing fruit which resembled grapes extremely fine; and its fragrance extended to a considerable distance."[19] Thence he proceeds to a minute view of beasts and birds varying in form as well as in the sounds they make.

Enoch was acutely observant; from heavenly mountain peaks he saw the entire universe of man and, from his guides and teachers, the angels Uriel, Raphael, and Michael, he learned the principles governing natural phenomena. They answered his questions and explained the "secrets of heaven," the regulations controlling the paths or courses of the sun, moon, and stars; they even allowed Enoch to see the sources and powers commanding the rain, winds, hail, and snow. In this heaven the jarring elements of frost or snow were not forbidden entry as in the Greek legend; here, instead, they had their origin and purposive explanation. Each element performed regularly at its proper periods without transgressing the orders received, and it was cooperatively related to other functions in nature. For example, the rivers and seas "together complete their respective operations," although the sea possesses a strong power which "causes it to ebb, so is it driven forwards and scattered against the mountains of the earth." A spirit constitutes the power in the sea as well as the frost, hail, or snow which has in it "a solitary spirit" that ascends from it like a vapor and is called "refrigeration."[20] Similarly, angels control the dew and rain in their close relationship:

> The spirit of dew has its abode in the extremities of heaven, in connection with the receptacle of rain; . . . the cloud produced by it, and the cloud of mist, become united; one gives to the other; and when the spirit of rain is in motion from its receptacle, angels come, and opening its receptacle, bring it forth.[21]

Natural laws were clearly animistic for Enoch. And the elements had their place in heaven. It even rained on the angels, but the quantity was regulated. Still it was a perfect heaven and the very embodiment of all knowledge.

Eight lengthy chapters of the book (71–78) are devoted to explaining the calendar, when the day was longer than the night, and how men greatly erred respecting the four extra days of the year because they did not understand how heaven's luminaries served in the mansion of the world: "one *day* in the first gate, one in the third gate, one in the fourth, and one in the sixth gate." These gates from which the planets issued forth were the exits of heaven, corresponding to the zodiac, and Enoch, with his countless statistics, showed himself to be an accurate but untutored observer of the heavens.[22] Repeatedly he mentioned the revolutions of the heavenly bodies to prove not only that the "harmony of the year becomes complete in every three hundred and sixty-fourth state of it," but also to prove that this was a purposive universe under God's command. At the close of the book he cautioned his race to be patient and await the plan of God.

It is apparent that Enoch's heavenly utopia retained little of the fragrant, misty, musical imagery of the Greek pattern, and his was far from a static scene. A dynamic force had made itself felt and a pervading sense of time moving from past to future became noticeable. Enoch literally moved through a succession of heavens, mountains, and valleys; he viewed panoramically the past in the Garden of Eden, the kingdoms to come for the just and unjust, and visualized actively the decisive day of last judgment. Professor Moses Hadas has suggested that "the Hebrew mode of thought is dynamic, the Greek static," so the contrast in heavenly concepts could be partially semantic in origin. " 'To sit' or 'to stand' in Hebrew are actually 'to move into a sitting or standing position.' " The Hebrew language has only the definite past and future tenses while the Greek has in addition the aorist tense denoting an action that took place in an unspecified past, which automatically introduces a timeless element. According to this theory then we find the Greeks generally concerned with space and symbols and the Hebrews with time and instrumentality, the process of construction and method of achievement. Although the comparison is overdrawn and generalized, as Dr. Hadas recognized, it is valuable in an area where precision is impossible.[23]

The contrast in attitude as seen in utopian writings of the Hebrews and Greeks may perhaps more safely be attributed to philosophic backgrounds of the authors—ethical and religious values of their respective societies. Let us consider Homer and Hesiod, religious exponents for their era as Enoch and Peter were for theirs, yet with points of view that vastly differ. Pagan and Christian writers part company abruptly in the underlying concepts of their heavenly utopias, regardless of the apparent similarities in adjectival description.

First of all, Homer and Hesiod were merely describing the locale, and its charms included the life of ease with material benefits understood; Enoch and Peter on the contrary were presenting transcendental heavens with emphasis on spiritual values. The last two were immensely concerned with the ethical life, good works, and righteous living that prepared a man for entry into heaven. In the Judaic-Christian tradition all stages of world history as well as a single individual's life were part of God's overall plan and providence, which gave man a sense of destiny in the world order, even as the Romans saw their history and development. The Hebrew or Christian was inextricably placed in this great scheme and required to believe, express faith, and humbly fulfill his small role in the cosmic plan. This constituted the method of achieving the blessed state and the rewards of immortality. Not so the older Greeks. Their elect were chosen on the basis of blood-relationship to the gods—as in Menelaus' case—or as heroes who carried out their individual fates proudly and courageously. Achilles was neither subjected to social responsibility for his race the way a Hebrew or an Aeneas was, nor was he expected to do more than remain true to himself. Philosophers and poets were gradually admitted to Elysium[24] but always on the basis of individual achievement and excellence, not because they had furthered Zeus' encompassing plans for the destiny of all human beings. Indeed, Olympian heights bred no such schemes, and Zeus himself executed the will of the Fates. Small wonder then that the Greek concept of bliss as seen by Homer, Hesiod, and Pindar was static, spatial, and limited only to a few chosen souls, whereas Peter and Enoch, reformers in attitude and passionate preachers who exhorted their followers to live righteously, offered rewards in a more dynamic heaven to a larger number of eligible candidates.

We have suggested that the contrast in outlook may have resulted from semantic differences or the enormous variance in philosophic belief between the two peoples. Yet there are always individual authors who do not bear the stamp of generalizations regarding their race and who caution us against too sweeping statements in such complicated areas. The two divergent views, for example, come closer together in the majestic Vision of Er reported by Plato in the conclusion to the Republic. In structure, in certain details, and in several concepts, the vision was a complete apocalypse, which may well have influenced Hebraic writers in the Hellenistic period[25] when Enoch lived. Thus it is possible that Greek thought as expressed by Plato in his best-known work played an important part in shaping later Hebraic visions. Plato's account of heaven and its operations bridged most successfully the gap between the older Greek concept and the more active Hebraic view.

On the Greek side, Plato's inheritance appeared in his avowed interest in the individual's soul, and even in heaven the individual was responsible for choosing his next life from the lots cast; moreover, Plato retained the powerful role of Fate, placing the three sisters prominently in heaven and assigning them their usual specific tasks; and he further maintained the position of Necessity both as mother of the Fates and the power turning the spheres of heaven. Yet, on what might be designated the Hebraic-Christian side, he showed the awards waiting for the just and the punishments in store for the unjust. Souls leaving their bodies journeyed to a mysterious region where there were two openings, an entrance and an exit side by side into the earth and two corresponding openings into the heavens. Here sat judges who sent the just to the right, upward into heaven, and the unjust along the road to the left and downward where they paid penalties tenfold for all the wrongs they had committed. Upon occasion the good souls, clean and pure, came down from heaven and joined those of the wicked who had expiated their crime and so were allowed to come forth though still full of squalor and dust. They met former acquaintances in a meadow with festival atmosphere and questioned each other on experiences above and below, thus revealing the contrast between suffering in hell and reaping reward in a heaven of delight and beauty beyond words. After seven days in the meadow they must travel on, view the workings of the circles of the universe (the third circle had the whitest color and the fourth was of a slightly ruddy hue), hear the musical harmony of the spheres, and, finally, from lots offered, they must select the form they will take in the next life—a selection that required a wise and knowledgeable man to differentiate the good from the bad.[26]

At the point of selecting lots, Semitic eschatology and Christianity clash with the Greek, but until now Plato's vision has not been very different from Peter's concept of heaven and the hell he afterward described. And the Vision of Er had the journey and general astronomical view Enoch later saw. Plato's thesis—that each of us should seek after and study the good, the ways in which it may be discerned, and reasons why the better life is one that is more just—could easily be adopted by Christianity and made to fit, even contribute to, its doctrinal writings, which, of course, is what happened centuries later.

Most important, however, is the fact that a progressive notion has entered into heavenly utopias both in the vision itself and the means of attainment. This dynamic approach clearly came in with the Judaic-Christian tradition though individual authors like Plato also indicated its presence. The concept of change and growth in a desirable direction was obviously vital if science was ever to gain a foothold in utopian kingdoms, for the idea of progress must precede or at least accompany it; development has to be possible in man's mind before he will devote himself to furthering knowledge and its applications. Alfred North Whitehead saw the basis for modern science in the ideas of a rational God and an orderly universe. Both these concepts appeared in the Hebrews' earliest utopian visions.

We must take care here not to sell the Greeks short or overvalue the Hebrews' contribution to the foundations for science in utopia. While Greek visions of the heavenly abode per se remained calm and settled—happiness in perpetuity—in reality their attitudes as expressed elsewhere by Homer and Hesiod were notable in encouraging man's investigation of the unknown. Although these attitudes prevailing in Greek life did not happen to enter their utopias, the gift to scientific thought was great indeed and reappeared with their legacy to the humanists in the Renaissance. Along with that Book of Fate, from which Zeus merely read and which Plato upheld as determining, was the strong and powerful individual, Achilles, Hector, or Ajax, making his own decision and determining his own fate. There were no puppets on the battleground at Troy though occasional mists made it difficult for the hero to see. Thus Ajax invoked Zeus while he challenged him: "Make heaven serene, and let us see; if you will that we perish, let us fall at any rate by daylight."[27] This is the spirit that endures in Greek thought, awarding greater admiration to Achilles than to Zeus himself, and resulting in a secular approach to knowledge. No priestly clan controlled manuscripts or experiment, and scientific thought in early Greece was consequently the work of laymen, independently reasoning, and defending their own opinions. Efforts of early thinkers to get at the nature of matter were further aided by the belief set forth in Hesiod's *Theogony*, namely, that this universe evolved from chaos. There was no religious theory of creation; instead, chaos produced earth which, in turn, by "exhalation or evaporation" created heaven. So there could be an infinity of worlds being born or dying. Most vital was this concept that celestial and terrestrial phenomena were essentially similar. This type of thinking was obviously of great value to the future of science, and we are faced with a paradoxical conclusion: the Greeks' actual contribution to scientific thought and approach—which did not emerge in their views of a better world—was greater by far than the Hebraic-Christian, though the dynamic, progressive element appeared actively in the latter's visions. The key lies in secularization and the lack of thought control. If "science is essentially an effort of man to help himself," as Benjamin Farrington defines it,[28] then the Greeks had greater freedom to do so; they were not inhibited by an omnipotent God, so that man's efforts did not seem futile or presumptuous.

Science in Heaven

The groundwork then may have been laid in heaven by the Hebrews for a dynamic approach to utopian states, but it was no more than fertile soil; no really scientific seeds had been planted. It is true that Enoch learned the "secrets of heaven"—the laws governing natural phenomena—but these "facts" were explained to him as a visitor in the realm above. Furthermore, he made it very clear that the knowledge is God's alone to reveal only if He so desires; it is not man's business to pry into His secrets or try to wheedle answers from Him. In other passages of *The Book of Enoch* we are told that Azâzêl (one of the fallen angels) was being severely punished, not just because he led the other delinquent angels who subsequently lay with mortal women, but because he and his friends

> taught men to make swords, knives, shields, breastplates, the fabrication of mirrors, and the workmanship of bracelets and ornaments, the use of paint, the beautifying of the eyebrows, *the use of* stones of every valuable and select kind; and of all sorts of dyes so that the world became altered.[29]

Other fallen angels taught sorcerers, observers of the stars, and astronomers, while Azâzêl especially "taught the motion of the moon." Destroyed by such learning, men cried out and, according to Enoch, their voices reached heaven, causing God's dire punishment to fall on the guilty teachers. Still they persisted and instructed men in writing and the use of ink and paper: "Therefore numerous have been those who have gone astray from every period of the world . . . for men were not born for this, thus with pen and ink to confirm their faith."[30]

Enoch's was an anti-intellectual attitude typical of religious reformers who wished primarily to instill faith in God and His benevolent providence. For many years to come, even as late as Milton's *Paradise Lost*, the issue of man's limits, his right or ability to know God's secret laws of the universe, was debated, and the foremost position was awarded to faith alone. Milton's Adam had learned this lesson and, as he was about to leave the Garden of Eden, admitted that he had had his fill "Of knowledge, that this Vessel can contain, / Beyond which was my folly to aspire." And the angel Michael commented:

> This having learnt, thou hast attaind the summe
> Of wisdome; hope no higher, though all the Starrs
> Thou knewst by name, and all th' ethereal Powers,
> All secrets of the deep, all Natures works,

Or works of God in Heav'n, Aire, Earth, or Sea,
. . . onely add
Deeds to thy knowledge answerable, add Faith.[31]

Thus Enoch taught in pre-Christian times, and future utopists in the 17th century had to wrestle with his argument (see Chapter VI) before science was freed to move ahead, invade God's realm, and learn His secret laws.

It may be observed with amusement, however, that applied science—automation—had entered heavenly utopias in ancient days. The miracle method was essential to provide all blessings without man's labor. And as early as Homer's *Iliad* there were golden maidservants to wait upon their master Hephaestus in his bronze palace among the gods. They looked like real girls, and Homer says: "In them is understanding at their hearts, in them are voice and strength, and they have skill of the immortal gods."[32] These girls appeared most valuable, and it was not surprising to find them in the household of an artificer like Hephaestus, who also created a set of three-legged tables on golden wheels "so that they could run by themselves to a meeting of the gods and amaze the company by running home again." The Greeks obviously had their inventive technicians (Daedalus was another) and the heaven of Olympus was not without automation. Nor was it without parody. In Attic comedy, Crates (fl. 450 B.C.) invented robots and equipment to care for food service. He ordered them: "Place yourself here, table! You, I mean, get yourself ready! . . . Where's the cup? Go and wash yourself. Walk this way, my barley-cake . . . Fish, get up!" "But I'm not yet done on the other side!" "Well, turn yourself over, won't you? and baste yourself with oil and salt." And the actor playing opposite continued the fun, adding that the hot bath would flow until ordered to stop, while the sponge and sandals advanced of their own accord.[33]

In the picture of the blessed heroes' life presented in the pseudo-Platonic *Axiochus* (371C), dated approximately the 1st century B.C., where poets and philosophers first entered and joined the old doughty warriors who held sway, the inhabitants were enjoying "agreeable banquets and feasts self-catered."

Parties produced and managed themselves—the dream of every hostess! A magical trick solved the problem of domestic service: no slaves or servants were needed and, of course, their entry into paradise would be difficult to justify when only the hero, the relative of the gods, and now the intellectually elite, were eligible.

Lucian too decided that the blessed should be spared every labor and provided in his *True History* for the automatic dispensation of fragrance.

> The heroes are even saved the trouble of putting on their own scent by the following ingenious system: specially absorbent clouds suck up perfume from the five hundred springs and from the river, after which they go and hover over the party; then the winds give them a gentle squeeze, and down comes the scent in a fine spray like dew.[34]

Lucian chose to explain the process. For most authors, it just happened in sheer miracle fashion such as manna falling from heaven or, as in Milton's jeweled heaven where the miracle was attributed to God's powers, the heavenly gates

> . . . self-opend wide
> On golden Hinges turning, as by work
> Divine the sov'ran Architect had fram'd.[35]

So the mechanical genius of the great artificer was not lacking in the Christian epic any more than it was in the pagan Homer's story.

Such details, however, are unimportant, small particulars in the great expanse of heavenly bliss; they are merely accessory and neither vital to the life of heroes, gods, or angels, nor essential to the concept of perfect happiness. No doubt the imaginative projection of automatic services helped to explain how the complete life of ease was possible (and we may, of course, recognize our present "seeing eye" door, dishwasher, catering service, and perhaps the "silent butler" or "dumb waiter") but, as envisioned in the early heavenly utopia or Milton's final statement of it, no relationship to scientific thought or development existed. Automation was magic and not even pseudoscientific.

Moreover, the exemplary heavens in which such clever services were available as in Homer, the *Axiochus*, Lucian, and Milton, generally reflected the timeless, more static way of life, not the dynamic. Bertrand Russell has drawn attention to Milton's static concept of heaven revealed in the lines from his poem, *At a Solemn Musick:*

> Where the bright Seraphim in burning row
> Their loud up-lifted Angel trumpets blow,
> And the Cherubick host in thousand quires
> Touch their immortal Harps of Golden wires,

And Russell humorously adds: "It is not suggested that the trumpets and harps should be of continually improved makes, or should be played by machinery to save the angels trouble and leave them free to increase the height of the buildings in the Golden City."[36] Although failing to observe that at least the gates were self-opening in *Paradise Lost*, he was quite correct in suggesting that our modern concept of progress had not invaded Milton's heaven. So Enoch's admonition to stay away from clever knowledge was echoed in religiously dominated utopias or ignored by those who did not even think of it seriously.

Within the religious context, utopian visions thus far have been scheduled for the future, whether located beyond the earth's rim like the Isles of the Blest and Elysian Fields or high in the skies like Peter's and Enoch's. The Messianic age of Amos and Isaiah was also to come in the future, but it was expected on this known earth. After all iniquity and corruption had been wiped out, God was to join his chosen race or the peoples of the world[37] and halcyon days were promised. For the early Judaic prophets there was no more thought of man's presuming to join God in His heaven than for the Greeks, who completely separated the gods' heaven from the blissful isle for heroes and half-gods. It was not until the apocalyptic writers (200 B.C.–A.D. 200) that it seemed conceivable for man to go into God's own kingdom.[38]

Regardless of location, all were visions projected for the future. And while they were evidence of man's desire for immortality, the wish to defeat death, which had haunted him since the days of Gilgamesh and probably before, they also amply provided for whatever was missing in the reality of his life and so expressed his goals, what he most desired, even if he had to await the hereafter for their realization.

In sharp and direct contrast to this hopeful attitude was the idea that the perfect age had passed and hence was irretrievable. From this point of view man had seen his best days and traveled a downhill course ever since; nothing better could be expected or worked for. Homer through the voice of Nestor (*Iliad* 1.260–268) had indicated the existence of a better people in the past, when men were strongest and conquered the wild ones who dwelt in the mountains; but he had not elaborated on the idea nor given the past age a name. It will be recalled that the Sumerians had united the perfection of early days and those to come for the blessed in the same place: Dilmun. But for us, it was not until the Garden of Eden and Golden Age became well known that we had a standard from the ancient past against which to evaluate our present state. Few facts concerning Adam and Eve's life in the Garden were given in Genesis; while many authors added imagery and various locations, the earthly paradise of our first parents received its richest description from Milton in his great epic. The Greeks' Golden Age was more completely presented by the early religious and ethical spokesman, Hesiod. According to him, those days in the dawning light of civilization were most pleasant for mortal men fashioned by the gods:

> These lived in the time of Kronos when he was king in Heaven. Like gods they lived, having a soul unknowing sorrow, apart from toil and travail. Neither were they subject to miserable eld, but ever the same in hand and foot, they took their pleasure in festival apart from all evil. And they died as overcome of sleep. All good things were theirs. The bounteous earth bare fruit for them of her own will, in plenty and without stint. And they in peace and quiet lived on their lands with many good things, rich in flocks and dear to the blessed gods.[39]

Subsequently the golden race became spirits by the will of Zeus, and they were the keepers of mortal men and moved cloaked in mist everywhere over the earth. Many years later Josephus (*Antiq.* 1.108) decided that Hesiod had given a life-span of 1000 years to people in the Golden Age and, in trying to explain their longevity, Josephus suggested not only that they were beloved of God and so enjoyed health-conducive diets, but that the long life was necessary "to promote the utility of their discoveries in astronomy and geometry. . . . For they could have predicted nothing with certainty had they not lived for six hundred years, that being the complete period of the great year." This is a bit of interesting interpolation from Josephus, though it has little relevance to Hesiod, who expressed no real concern with natural philosophy or scientific developments in his writings.

More important for present purposes is the fact that Hesiod saw the world declining after the Golden Age, through the Silver, Brazen, and Heroic Race, which had its Isles of the Blest, to the Iron Race of Hesiod's own time. According to this concept of history in which the best periods have passed eons ago, we have neither hope for betterment nor interest in achievement: man has deteriorated as imperfection in the present age proves. This pessimistic view leads to Horace's degenerate sires begetting more degenerate sons, and it was strengthened along the way by the words of the Psalmist assuring us that the world "shall wax old like a garment." Within such a depressing framework, no ideas of progress take root, no goals flourish; the utopian dream is over and finished and life is steadily, irrevocably worsening. Fortunately this negative outlook, though always with us, has never held for long and it has been actively opposed by some optimistic souls in every generation. Even while the Greeks accepted the idea of decline, they held also (and one feels more strongly) the opposing, constructive belief of the Prometheus myth that man had started in a primitive state and benefited greatly from Prometheus' daring act when he stole the fire from heaven and gave man his first means toward civilization. From that time man had advanced in knowledge and competency in dealing with the unfriendly elements. This belief in progress helps explain the Greeks' attitude of superiority over all other peoples, foreigners or barbarians. Although they recognized no Christian providence, but seemed to accept the rule of the three sisters of Fate, their confidence in themselves as capable individuals has seldom been excelled. And the prophets of the Old Testament may have ranted against the wickedness in which they lived and which had to be destroyed before the regenerated race could enjoy the bounteous era, yet they promised the Messianic kingdom to come after the judgment. Beneath their belief, of course, was their never-failing faith in God's providential management of human affairs.

Thus it would appear that whether rooted in the more secular thought of the Greeks or the Hebrews' religiously oriented doctrine, the idea of society's progressive development was somehow supported, and it existed at least alongside the defeatist concept of historical decline. The belief in possible improvement was the ancients' greatest bequest to future utopists and supplied them with philosophic foundations on which they could construct their ideal states. Within this concept of human worth and destiny the Hebrews' dynamic view invaded the static utopia projected generally by the Greeks. Still the aim of all utopian hopes for the future was moral virtue. It was toward this goal that man's progress was directed and only ethical behavior, whether according to Judaic or Greek standards, could prepare him for entry into the heavenly paradise.

II

SECULAR VIEWS OF UTOPIA

If the heavenly gates were restrictive to utopists' imaginations because of theological doctrine and time limitations (paradise was in the future or dim past), the restraints vanished when the new city was set on earth. Relocated, it could exist even in the present time and be reported upon by travelers and others who often furnished the "facts" to historians. This freed poets to embellish and beautify the ideal environment; biographers could describe the ideal life of a human being for others to emulate; teachers might submit plans for the perfect educational program; and philosophers could project the society regulated in its many aspects by their own systems of thought. Unlimited visions could explore numerous ways to achieve the better world through a variety of methods. New elements thus joined and contributed to the growth of idealistic worlds. During this early period of classical brilliance there appeared the first full-scale utopia, the forefather and archetype of the literary genre as it is known—the *Republic* of Plato.

It is essential, however, to see his predecessors as well as his imitators who appeared soon after his popular *Republic* was known; too often the story of utopia has assumed that the *Republic* miraculously gave birth to the genre, even as Athena sprang full-blown from the forehead of Zeus. Ethical principles, of course, persisted as a major concern for Plato and many other visionaries, but without necessarily promising the rewards of immortality and heavenly bliss. Religious overtones might be heard or implied but no longer did eschatological notes dominate the harmony of the scene.

The Ideal of Nature

Yet, heavenly visions of the perfect land bequeathed their descriptive pattern, surface imagery, to secular views, particularly those laid in gardens and pastoral settings. The grain-giving earth and abundant crops, the luscious vineyards, flowers and fruits moved easily from Elysian Fields, Isles of the Blest, the Golden Age, and Amos' Messianic vision into the nontheological garden sometimes intended for young, beautiful lovers or the pastoral setting for a shepherd's life of pure simplicity, close to nature under friendly heavens. Adjectives and metaphor might be the same, but the purpose was very different. For example, in Greek romance Longus describes the garden of Philetas where he had encountered Eros and received word that he should instruct Daphnis and Chloe in the ways of love. Certainly there was no more enticing beauty in a heavenly utopia than in Philetas' garden, made by his own hands:

> Whatever the seasons bring my garden produces. In the spring it has roses, lilies, and hyacinth, and both kinds of violets; in the summer poppies and pears and all varieties of apple; now it has vines and figs and pomegranates and green myrtles. To this garden troops of birds make their way together each morning, some for food and some to sing, for it is overarched and shady and abundantly watered by three springs; if one would remove its hedge he would fancy he was looking at a natural wood.
>
> When I entered my garden about noon today I espied a little boy under my pomegranates and myrtles, some of which he was holding in his hands. His complexion was as white as milk, his hair bright as fire, and he shone as if he had just been bathing.[1]

The imagery—even the red and white coloring—has been transferred with little change from heavenly to earthly visions of utopia.

But notice that in the 3rd century A.D., when Longus was probably narrating his romance, there was a studied artificiality behind the apparent simplicity. Philetas' greatest compliment to his garden was that if the hedge were removed, one would "fancy he was looking at a natural wood." His garden, artificially created, imitated the true beauty of nature herself. Furthermore, Longus chose for his spokesman an elder statesman of bucolic poetry in Philetas (born not later than 320 B.C.), whose pastoral elegies had become models for the literary form and were so acknowledged even by Theocritus who was his junior by only a few years and who devoted his *Idyll* 7 to Philetas. The pastoral setting re-created by Theocritus, a city-dweller, was also artificial but a sincerity of tone belied it. His *Idylls* had a rare atmosphere of beauty and charm, a true feeling of the primitive and simple shepherd's life: his song, his pipes, the oak tree and hyacinths, his love, his friends, and the flocks he tended under a wide expanse of blue skies. His tables were not piled high for banquets and he had to make his own music, neither automatic service nor an angelic choir existed. Yet his life had a peaceful constancy, the Arcadian setting that appealed so strongly to the urban imagination, harassed by violence, noise, and change. Vergil adopted this setting for his *Eclogues* but the artificiality became more apparent and the scene served as a mere backdrop for personal and political comment. Still the patterned imagery continued, and in skillful hands it emerged both in romantic bowers of bliss, the luscious temptation scenes in Renaissance poetry, and in the quite opposite return-to-nature movement, the ideal of simplicity unspoiled by sophistication and materialism.

Of all preceding utopian concepts the Golden Age, first pictured by Hesiod, contributed most directly to the ideal of nature as it was in the beginning, when man was beautiful and innocent. Major adjustments to Hesiod's early view, however, had been made by the Stoic Aratus[2] (c. 315–240/239 B.C.), who removed supernatural traits and presumed that men might regain the good, primitive life if they were so inclined. The simple life for the Golden Ones seen by Aratus was a "hard" primitivism compared to the earlier life of easy bliss, the "soft" primitivism allowed by Hesiod.[3] Aratus stressed the presence of Justice on this earth during the Golden Period, which knew neither hatred nor war. There was a noticeable absence of foreign trade, and men were vegetarians; the oxen drew the plough as they were so intended, and it was not until the ensuing Bronze Age that man forged the sword and killed the beasts for food. The contamination of civilization was spreading. To this statement Ovid added the finishing touches and described a Golden Age which was the model for years to come; his Golden Age returned to the idyllic ease of the earliest picture and included all the magnificent imagery.

> The first age was golden. In it faith and righteousness were cherished by men of their own free will without judges or laws. Penalties and fears there were none, nor were threatening words inscribed on unchanging bronze; nor did the suppliant crowd fear the words of its judge, but they were safe without protectors. Not yet did the pine cut from its mountain tops descend into the flowing waters to visit foreign lands, nor did deep trenches gird the town, nor were there straight trumpets, nor horns of twisted brass, nor helmets, nor swords. Without the use of soldiers the peoples in safety enjoyed their sweet repose. Earth herself, unburdened and untouched by the hoe and unwounded by the ploughshare, gave all things freely. And content with foods produced without constraint, they gathered the fruit of the arbute tree and mountain berries and cornel berries and blackberries clinging to the prickly bramble-thickets, and acorns which had fallen from the broad tree of Jupiter. Spring was eternal, and the placid Zephyrs with warm breezes lightly touched the flowers, born without seeds; untilled the earth bore its fruits and the unploughed field grew hoary with heavy ears of wheat. Rivers of milk and rivers of nectar flowed, and yellow honey dripped from the green oaks.[4]

Nothing but Milton's Garden of Eden in its pristine state can match this abundancy. Consequently, instead of attempting additions, later idealists simply called for a return to primitive perfection and adapted that concept to fit their own age, which was afflicted with materialism and false values. With this attitude Montaigne commended the primitive, simple life in his essay *On*

Cannibals, and Rousseau saw the value of the noble savage freed from diseases infecting the polished, overly refined life in Europe's capital cities of the early 19th century. Among more recent expressions of the ideal of nature is W. H. Hudson's *Crystal Age*, which presents a glistening vision of the beautiful peaceful life: the utter simplicity of human relationships, close to the animals and nature, enchanting music of the human voice, and the complete absence of strife of any sort, even passionate love.

Although this ideal never fails to charm the human race and thus endures, it is at heart a withdrawal, a negation of the concept of progress in the sense of forward movement. It asks us to return the clock back and regain the Golden Age; scientific research for advancement is an anomaly and, in later utopias like the *Crystal Age*, such efforts were definitely repudiated. Instead, handicrafts and agriculture formed the economic basis for the society, and man created his own food, clothing, and necessities as well as his amusement and entertainment. If not a complete reversal of progress, it was at least a different goal, which required a rather thorough demolition of modern life and its accoutrements, the wholesale removal of equipment, machinery, technical developments, monetary standards and all accompanying values. Hudson realized this and revealed the impossibility of modern man's adjustment to the simple life of nature by placing a Londoner from the fashionable West End, wearing his well-tailored suit and boots, in the devastating position of trying to get along with natural people to whom he seemed a ridiculous, coarse boor. Needless to say, the poor chap suffered and finally failed. The ideal of nature triumphed once more and in varying degrees of emphasis it has endured in many utopias since it was first projected in the Golden Age, classical gardens, and the pastoral countryside.

The Ideal People

The idealistic concept of a perfect people was more viable for utopias than the ideal of nature, which could not long be sustained with vitality as civilization developed. An ideal people immediately involved society at large, all the complicated relationships among citizens, workers, families, the government, and laws. For this reason ancient stories of ideal people come closer to full-scale utopias, though often only a fragment remains extant. From these, it is a main line to future, more developed utopias describing many aspects of a complex life.

Ethiopians appeared to be such an ideal people, unusually fortunate perhaps because their land was watered by one of the four rivers issuing from the Garden of Eden (Genesis 2.13). Or maybe Herodotus correctly explained their felicitous location, though he doubted his own story in this instance and made clear that he was only repeating it as told to him: namely, that a fountain too deep to measure (an Egyptian king had tried lowering a rope many thousand fathoms in length) fed half its waters to the great Nile River and gave the other half to Ethiopia toward the south.[5] Regardless of the source of its water supply, the land was lovely, green, and fertile. Homer established it at the rising or setting of Hyperion on the stream of Oceanus and related that Menelaus visited there on his journeys and found that

> the lambs have horns as soon as they are born, and the sheep lamb down three times a year. Every one in that country, whether master or man, has plenty of cheese, meat, and good milk, for the ewes yield all the year round.[6]

Ethiopians were a milk-drinking people, like the righteous Aboi admired by Zeus, and ate only boiled flesh, to which diet they attributed their long lives; most of them lived to 120 years and some beyond. In fact, their king had never tasted wine until it was presented to him among gifts from the Persian king Cambyses; then he enjoyed it thoroughly and decided it must be the palm wine that accounted for the Persians' 80 years of life. He liked the wine so much he confessed that in this drink at least "the Persians surpassed the Ethiopians." While explaining his own people's longevity, the king led his guests to a fountain wherein they washed and

> found their flesh all glossy and sleek, as if they had bathed in oil—and a scent came from the spring like that of violets. The

> water was so weak, they said, that nothing would float in it, neither wood, nor any lighter substance, but all went to the bottom.[7]

And Herodotus added, "If the account of this fountain be true it would be their constant use of the water from it which makes them so long-lived." Or it may be explained by the existence of their "table of the Sun" which was a sight King Cambyses was most eager to have his ambassadors see. The account of it given to Herodotus was thus:

> It is a meadow in the skirts of their city full of the boiled flesh of all manner of beasts, which the magistrates are careful to store with meat every night, and where whoever likes may come and eat during the day. The people of the land say that the earth itself brings forth the food.[8]

So there was plenty for all and freely given to prince and shepherd alike as Menelaus had claimed.

Moreover, they were an extremely sensible people: the king laughed at a gold neck-chain and armlets among his gifts, "fancying they were fetters" because Ethiopians used golden chains on prisoners and he claimed theirs were much stronger than the Persians'. (Of all metals copper was the most scarce and valuable in the land.) In other policies Ethiopians showed themselves equally wise as, for example, when the king calmly advised the emissaries (really spies from Cambyses) to tell their king he was not a just man to covet the land of others who had done him no wrong, and he sent a bow of great strength suggesting that when the Persians could pull it easily, then let them come against the long-lived Ethiopians. Fortunately, when Cambyses, angered by this charge, set forth to attack, he made the fatal mistake of failing to provide for his troops and so met disaster, thus leaving the ideal society of Ethiopians intact. As Herodotus summarized it:

> Where the south declines towards the setting sun lies the country called Ethiopia, the last inhabited land in that direction. There gold is obtained in great plenty, huge elephants

> abound, with wild trees of all sorts, and ebony; and the men are taller, handsomer, and longer lived than anywhere else.[9]

These people are certainly worthy of emulation, which may account for Herodotus' elevation of obviously mythological traits to the level of historical reality. Here Heliodorus will lay the final scenes for his adventurous romance, *Aethiopica.* Years later, Ariosto, during the Italian Renaissance, will imaginatively place the Garden of Eden on a mountaintop in Ethiopia and permit his heroic knight Astolpho to taste the fruit, so delicious he concludes:

> Our first two parents were to be excus'd,
> That for such fruit obedience they refus'd.

And from this fabulous garden Astolpho takes off for the Valley of the Moon where he recovers Orlando's lost wits.[10]

Undoubtedly Sir Thomas More remembered that Ethiopians used gold for fetters when he too made it into "great chains, fetters, and gyves wherein they tie their bondmen," and he added its use in making chamber pots and children's toys.[11] Furthermore when ambassadors wearing armlets and chains of gold arrived in Utopia just as they had in Herodotus' account of Ethiopia, the Utopians, completely misunderstanding, thought them court jesters or fools, and children seeing bright jewels "sticking upon the ambassadors' caps, nudged their mothers and said to them: 'Look, mother, how great a lubber doth yet wear pearls and precious stones as though he were a little child still.' "[12] So the trend continued in utopias and finally Voltaire paved the streets of Eldorado with gold, made quoits of it and other useless metals for children to toss, and his natives burst their sides with laughter when the visiting Candide tried to pay with gold.[13]

Noteworthy too is the fact that Ethiopia was at the edge of the world on the maps of that time, for the idea grew more easily into tradition and provided an appropriate location even for a 17th-century utopist like Johann Andreae, who survived a shipwreck in the Aethiopian Sea and found nearby his perfect land. It may have been for the best that Herodotus never actually got there; he only traveled as far as Elephantiné, then he faithfully stated that his report was from "inquiries concerning the parts beyond." Certainly we can never know whether Homer really saw the Ethiopians, whose name was often used indiscriminately for any dark-skinned people living far away on the southern border. Thus, although other Greek travelers actually did visit there as early as 665 B.C., these imaginative accounts helped the story to grow and the Ethiopians to become a legend in ancient days.[14]

Competing with the blameless Ethiopians of the south were many others scattered about in outlying districts of the world where grass could always be greener. In the northern region bordering on *Terra Incognita* lived the legendary race of Hyperboreans,[15] highly regarded by the Greeks, since they were devout worshipers of Apollo and, according to Delphic tradition, the god spent his winters with them, even as Zeus passed an occasional weekend with the exceptional Ethiopians. These comparable peoples of the torrid and frigid zones were deemed worthy of the gods' company and, for this reason perhaps, one dared not look too closely into specific details of their lives. Consequently, their remarkable feats were generally shrouded in impenetrable and protective clouds. The Hyperboreans simply had marvelous and mysterious means for transporting their gifts to the Delian Shrine from their faraway land and, though there was speculation on who did it and what route they followed,[16] the answer was always an enigma; thus Herodotus expressed skepticism regarding their existence and recounted "tales" concerning them, concluding that "if there are Hyperborean men, there are others too who are Hypernotions."

Even their name bears the vague meaning of "beyond the north wind" or "beyond the mountains." And when Pindar sang their praises, the secret of their whereabouts remained in obscurity:[17] "Neither by ships nor by land canst thou find the wondrous road to the trysting-place of the Hyperboreans." Nevertheless, Pindar assured us, in olden days Perseus had found them, watched their honorable sacrifices, and joined them in banquet festivities at which Apollo enjoyed himself, laughing happily at one entertainment provided by "brute beasts in their rampant lewdness." Yet this crude performance in no way implied lack of culture among Hyperboreans:

> Such are their ways that the Muse is not banished, but, on every side, the dances of maidens and the sounds of the lyre and the notes of the flute are ever circling; and, with their hair crowned with golden bay-leaves, they hold glad revelry; and neither sickness nor baneful eld mingleth among that chosen people; but, aloof from toil and conflict, they dwell afar from the wrath of Nemesis. To that host of happy men, went of old the son of Danaë, breathing boldness of spirit, with Athena for his guide. And he slew the Gorgon, and came back with her head that glittered with serpent-locks, to slay the islanders by turning them into stone. But, as for me, in the handiwork of the gods, nothing ever seemeth too incredible for wonder.[18]

To Pliny also the incredible qualities of Hyperboreans were understandable and acceptable as the handiwork of the gods; he did not doubt with Herodotus, and as the legend grew, its credibility was upheld. In his *Natural History*, Pliny included them with known tribes,[19] and the story passed into medieval literature with belief. Often the Scythians of the north—a real primitive people—were associated with Hyperboreans, apparently their near neighbors or inhabiting the same general geographic area, but the Scythians did not emerge in unadulterated glory as the ideal. Contrary to the Ethiopians, also of course real people, the Scythians' customs were not approved by Herodotus, nor were they, in his opinion, wise men. Hence, though they possessed extraordinary qualities mingled with the realistic, they do not belong with the utopian literature.

On the western boundary of the world, however, only five days' run from Britain, there was a genuinely strange and highly imaginative people: the Ogygians who dwelt in such abundance they could spend their time in discourse and philosophy. Such intellectual pursuits were unique among fortunate peoples generally, but then the Greeks had direct influence on the island of Ogygia. Plutarch reported[20] that here Cronus lay pleasantly sleeping though confined in a deep cave of rock that shone like gold and his dreams were revealed as prophecies. Here too Heracles left some of his men who "rekindled again to a strong, high flame the Hellenic spark." Visitors to this rare spot were required to stay for 30 years, the time required for the star of Cronus to complete its cycle and enter the sign of the Bull, most propitious for sacrifice, after which they were allowed to sail home. But most chose to remain, because all things were plentiful without toil, the island was suffused with fragrance, and there was a marvelous softness in the air. More rare was the fact that one stranger visiting here explained that he became "acquainted with astronomy, in which he made as much progress as one can by practising geometry, and with the rest of philosophy by dealing with so much of it as is possible for the natural philosopher."[21] Perhaps because of this unusual feature of Ogygian life, Johann Kepler became so fascinated with Plutarch's essay that he compiled an annotated edition and tried to identify each island of the group.[22] In doing this he had inevitable trouble, because Ogygia was identified as Calypso's Isle and, owing to Cronus' presence, was easily confused with the Fortunate Islands, which from time to time were also associated with Islands of the Blest and Elysium. If Ogygia was Iceland (Island), as has also been claimed, one only wonders what happened to the Ogygians' early lead in scientific studies, their single most distinguishing accomplishment.

Farther to the south but still on the western limit of the map was the fabulous continent of Atlantis about which Plato wrote. Following a hypothetical discussion of an ideal republic, Socrates had suggested that it might be wise to evaluate the effectiveness of such a state in time of war or great emergency to determine whether the ideal society could be successful in performance. So on the next day when he and his friends gathered to continue their feast of discourse, it was agreed to transfer the fiction of the *Republic* "to the world of reality," as Socrates put it, and consider it to have been ancient Athens, a truly remarkable society and most successful in its great military action against the race of Atlantis. Now Critias, who had told the story of primeval Athens[23] (handed down in his family from his great-grandfather, a dear friend and relative of the lawgiver

Solon, who had learned it from Egyptian priests), was to go on and relate also what he had heard of their formidable opponents on Atlantis.

Thus he explained the Atlanteans' descent from Poseidon, who had received this large island outside the Pillars of Hercules when the gods divided the region of the earth in the very beginning. There, with Clito, the only daughter of earth-born parents who conveniently died when she was "husband-high," Poseidon produced five sets of male twins and allotted the land among the ten princes. The eldest was Atlas, whose name was used to designate both the Atlantic ocean and the race of people on these shores. Hence Atlanteans were semi-divine in origin, and the miraculous nature of their territory and development was partially explained.

One of the initial acts of Poseidon was to fortify the hill in the center of the island where Clito lived by fashioning a series of circles, alternate rings, three of sea and two of earth, getting successively smaller toward the center and making the spot inaccessible. The heart of the island "he adorned with his own hand—a light enough task for a god—causing two fountains to flow from underground springs, one warm, the other cold, and the soil to send up abundance of food plants of all kinds." With such wealth the island was self-sufficient, yielding "all products of the miner's industry, solid and fusible alike . . . a generous supply of all timbers serviceable to the carpenter and builder," and ample lands for the pasture and maintenance of many wild or domestic animals, including the mammoth elephant.

> Besides all this, the soil bore all aromatic substances still to be found on earth, roots, stalks, canes, gums exuded by flowers and fruits, and they throve on it. Then, as for cultivated fruits, the dry sort which is meant to be our food supply and those others we use as solid nutriment—we call the various kinds pulse—as well as the woodland kind which gives us meat and drink and oil together, the fruit of trees that ministers to our pleasure and merriment and is so hard to preserve, and that we serve as welcome dessert to a jaded man to charm away his satiety—all these were produced by that sacred island, which

then lay open to the sun, in marvelous beauty and inexhaustible profusion.[24]

But Atlanteans did not rest content in plenitude. Soon they had developed a great empire extending to Tyrrhenia (southwest Italy) and Egypt, thereby adding external revenues to their original riches. They built an elaborate system of bridges and canals through plains to transport produce and lumber from the forests to the port and to provide irrigation for fields, forcing a second crop annually; huge palaces were constructed, a temple to Poseidon with statues of gold on a tremendous scale; a race track, gymnasia, dockyards and harbor, areas for the military and war equipment—all were developed to make this one of the most thriving and powerful civilizations ever known. To govern their territories, the ten princes met on specified occasions (alternate intervals of four and five years, "thus showing their respect for even numbers and odd") and deliberated on common affairs, inquired if any had transgressed the law, and passed judgment with proper ceremony and ritual.

For many generations these people were both true and great-hearted, obedient to their laws, valuing virtue above all else, and conducting themselves with fair judgment and humility toward each other. The weight of their gold and other possessions was light: "Wealth made them not drunken with wantonness; their mastery of themselves was not lost, nor their steps made uncertain." This state of wisdom endured as long as the god's strain in them was vigorous but when, by constant mixing with mortals, the human temper became dominant, they began to behave in unseemly fashion with pride and ambition for power. So the gods took notice and at the end of the fragmentary *Critias* we are led to believe that for loss of virtue, the civilization of Atlantis fell, first to the Athenian forces and then fatally in an earthquake that engulfed the entire continent and its inhabitants. Only the memory remained, but it was a lasting memory from Plato's record on through the Middle Ages and into the 17th century, when Francis Bacon named his fable *New Atlantis.* During the years its unfortunate moral decadence and final collapse were forgotten, and it became a symbol for a perfect people and a happy land. Here the Garden of Eden has also been placed, in addition to its many other locations; here it has been claimed that the first root race for all mankind arose; and numerous theories about its actual existence have led many men, responsible like the Prime Minister of England, William Gladstone, as well as irresponsible cranks, to promote projects[25] for revealing the former beautiful civilization now lying in the depths of the Atlantic Ocean. Atlantis was an admirable place in its day and has served as an expression of imaginative utopianism ever since Plato told the story.

On the southeastern boundary of the ancients' flat world there were two other famous utopian societies with secular characteristics: Iambulus' Islands of the Sun (2nd century B.C.) and Euhemerus' account of the Panchaeans (c. 300 B.C.). Both of them reflect the older wonderland of Atlantis and features of Plato's *Republic* while they also contribute to the growing structure of the future utopia. Euhemerus[26] of Messene was requested by King Cassander, whom he served from 311 to 298 B.C., to make a long journey abroad, which led him from the coast of Araby the Blest (so called for its great fertility) out into the ocean for many days until he arrived at the shore of some islands, one of which bore the name of Panchaea. Here he saw a rare people deserving of a place in history's record, for they excelled in piety and honored the gods "with the most magnificent sacrifices and votive offerings of silver and gold." This island was singularly blessed because, according to an old myth, Uranus and Zeus had dwelt there and been man's benefactors before going to their heavenly abode, just as Atlanteans had been privileged people, at least for a time, as descendants of Poseidon.

Furthermore, the inhabitants of Panchaea were as advanced in development and as wealthy in natural resources as their predecessors. Around the temple of Zeus Triphylius[27] was a plain "thickly covered with trees of every kind, not only such as bear fruit, but those also which possess the power of pleasing the eye." There were enormous cypresses, plane trees, sweet-bay, and myrtle; meadows were teeming with varied plants and flowers; birds in lofty forests delighted men with their beautiful songs; grapevines of the usual great number and of every variety climbed high and intertwined so as to charm and delight the viewer; fresh water sprang forth, forming a river for boats and irrigation, and at its source the water was clear, sweet, and conducive for the health of the body, so the river bore the name "Water of the Sun." Above the plentiful plain lay the mountain where Uranus used to look at the heavens and its stars, and beyond this was a multitude of wild beasts: elephants, lions, leopards, and gazelles. It was the typical utopian setting.

Euhemerus went on to explain the political organization of the Panchaeans, which was not like the Atlantean monarchy, but was reminiscent of Plato's aristocratic division in the *Republic*. The people were divided into three castes:

> The first caste among them is that of the priests, to whom are assigned the artisans, the second consists of the farmers, and the third is that of the soldiers, to whom are added the herdsmen. The priests served as the leaders in all things, rendering the decisions in legal disputes and possessing final authority in all other affairs which concerned the community.[28]

Farmers received rewards of first, second, and on to the tenth place for special produce, so competition was encouraged. Yet no man owned anything for himself except his house and his private garden. All products and revenues were collected by the priests and distributed with justice to each man. Still luxury existed: the people were clothed in softest wool and wore golden ornaments like the Persians; priests were more elegantly dressed in robes of sheer linen or even finer wool, in elaborate headdresses of gold, and in sandals of varied colors. They lived with every comfort, far above the other castes. Thus the society of Panchaea was religiously centered and under control of the priesthood, but it was also a thriving economic community based both on a system of incentives or rewards and on a community of goods that resulted apparently in a happy people living well.

Comparatively, Iambulus' Islanders[29] were no less content, though circumstances were a bit different; life was more communal, marriage was unknown and women were held in common, and children were exchanged, so that no one knew his own son. Meals were not served communally, though they were prescribed for certain days, that is, fish, fowl, or meat, in order to keep a healthy people free from disease. Anyone with a defect was expected to take his own life. Society was organized in clans of no more than 400, and each group was ruled by the eldest member, who also obligingly put an end to his life at age 150 so that the next in years might succeed. Insofar as we know from the fragmentary record of these people, their life was less dominated by religion than the Panchaeans'. Here the gods were the all-encompassing heavens and celestial bodies with the sun foremost in worship, hence the name, *Islands of the Sun.* The main island seems to have been conceived as a microcosmic reflection of the great sun above: it too was circular in shape and enjoyed day and night equally. It is also understandable, with their devotion to heavenly bodies, that astrology should be their chief concern though every branch of learning was encouraged.

The circular form—evidence of complete perfection—was seen earlier in Plato's description of Atlantis and it will be encountered frequently in later utopias. Its dominance in Campanella's *City of the Sun* in the 17th century was most apparent: the whole city was circular in design with seven concentric walls, and the similarities extended further, since Campanella's citizens also worshiped the sun and inordinately valued the study of astrology. It is possible that Campanella read the Italian translation of Iambulus made by Ramusio[30] in the 16th century, but direct influence is unimportant in this case; the circular plan was prevalent in many ideal designs because it connoted the perfect.

Iambulus contributed in other ways as well to the growing designs for utopia: first, he selected the choice location on the equator, which was not considered inhabitable until the great geographer Eratosthenes established the idea that it was.[31] Secondly, he made extensive use of various techniques to authenticate his narrative, to make it believable and real. His abundant scientific vocabulary, including botanical, anthropological, zoological, astronomical, and geographical terms, was so convincing that, as in the case of Atlantis and Ogygia, subsequent efforts were made to prove the exact location of the Islands of the Sun. As late as 1944 two interested writers identified the main island: one held it was Madagascar and the other decided it was in the Indonesian Archipelago.[32] We are a credulous human race. This is exactly the reaction the early utopists wished to provoke, and it verifies the success of their techniques.

From the isle of Panchaea it had been claimed that one could descry India, only a bit "misty because of its great distance," and there had certainly been little doubt about where the Ethiopians lived; at least it was in the southernmost region. But while the ancient authors established geographic locations in terms of recognizable lands they, at the same time, carefully selected the edges of the earth—south, north, east, and west—places few in their audience could ever have seen. Often they chose an island, which was a natural choice because it was a haven for seafaring people who knew the joy and relief of reaching a small piece of land after a threatening, storm-tossed sea.[33]

In addition to the bequest of locations to future utopias, the element of authenticity has received further endorsement from the early secular writers. It will be recalled that apocalyptic writers like Enoch, describing religious heavens, had also revealed this concern and tried to create the feeling of reality by giving many specific details to counter the fact that their source was completely visionary. Now, in secular views of an ideal people more ways have been devised to make the reader accept the whole account as true. When a historian like Herodotus claims a reliable informant and tells about Ethiopians or Euhemerus who traveled to Panchaea records an eyewitness account, the tale sounds true.

Although many enjoy and expect a dash of exaggeration in the traveler's story of strange, faraway places, the realistic tone of presentation has thoroughly annoyed some readers. Strabo commented on the tendency of travelers to wander from the truth in their reports;[34] and Lucian directly opposed them when he titled his work *The True History* and stated his satirical intention: every episode was to be a parody "of some fantastic 'historical fact' recorded by an ancient poet, historian, or philosopher." Among those singled out by Lucian to receive his sardonic barbs was Iambulus,

> who told us a lot of surprising things. . . . They were obviously quite untrue, but no one could deny that they made a very good story, so hundreds of people followed his example and wrote so-called histories of their travels describing all the huge monsters, and savage tribes, and extraordinary ways of life that they had come across in foreign parts.[35]

He continued his diatribe, naming "the real pioneer in this type of tomfoolery," Homer's Odysseus, who told "Alcinous and his court an extremely tall story about bags full of wind," evidently thinking his audience "fools enough to believe anything." Discredit the tall tales as you will, they continue to enthrall and carry the reader to limits of credulity with the reassurance of a pretended actual base. Utopias employ this combination of factual approach to the imaginative world just as romantic fiction has used the same paradoxical union.

Furthermore, these stories of ideal people, whether written by historians, poets, philosophers, or essayists like Plutarch, have considerably enlarged the content area and some might be called full-fledged utopias if more information had survived. As it is, an increasing number of the aspects of society are described: politics, economics, social relationships, and programs of learning. Although little is known of Hyperboreans and Ogygians, considerable detail remains regarding Ethiopians, Atlanteans, Panchaeans, and the inhabitants of the *Islands of the Sun.* In these last-named societies, political organization was authoritarian, dominated by the monarchy, priesthood, or rule of the eldest; economic and social systems ranged from control by the few to more communal living as among Iambulus' Islanders. In all, however, justice and equity were claimed, and virtue was understood to be the highest goal. There was no question of further progress or development in these societies, their present operative plan was successful in superlative degree. Only in the *Islands of the Sun* and Ogygia was the question of learning raised, and in the latter it was a rare curiosity to find scientific studies pursued with diligence and obvious seriousness. Thirty years were required for their mastery, because one had to stay long enough for the heavens to complete a cycle and to prove the observer's calculations and predictions right or wrong. Still there was no suggestion as to what Ogygians learned in their scientific research or whether they applied its results to provide a better life for their citizenry.

In the various views of ideal people, which clearly were popular reading in the days of the ancients, we see the early Greek interest in strange, extraordinary peoples as well as their concern for improving the social order. Stories of remote civilizations could be and were used to imply criticism of currently accepted practices and to suggest better ways of doing things. With this development the utopia approaches the form as we know it from Plato onward and indeed his *Republic* may well have given Iambulus his ideas of communal living and Euhemerus the basis for his three caste socioeconomic system. But one other important ideal requires attention before all lines merge in the famous *Republic*.

The Ideal Individual

In the Greek mind nothing was more important than the development of an individual man, bodily, mentally, and spiritually. His potential was immense and the responsibility his own; glory and fame were his rewards and constituted his "immortality" without necessarily guaranteeing entry to the Isles of the Blest. Man on this earth was sufficiently vital to be "the measure of all things" as well as an object of supreme beauty. A perfectly sculptured human body was placed in the town square for all to see every day; it was as unattainable in its beauty on a high pedestal as if placed in heaven or in ideal societies on the world's periphery. Still it served a function for the Greeks in its superhuman beauty: it exemplified perfection carved with tutored skill and great effort from imperfect, raw materials. So also could imperfect human nature be skillfully led and taught to reach the heights of mortal perfection and excellence.

Examples set by illustrious men in lawgiving, speaking, and noble acts were held up before the public as worthy of emulation. Biographers recorded great lives as the epic poet immortalized renowned deeds and the names of heroes. But of all means used to achieve the statuesque goal, education was foremost in Greek emphasis, and to it may be attributed not only their distinction in leadership and performance in the great classical period, but also their remarkable influence in the Hellenistic age when, as a minority without military power, they literally dominated the majority, the countries of the Mediterranean world and beyond, through the establishment of educational centers.[36]

Philosophers who were teachers took it for their purpose to fashion the ideal gentleman. Of course, their ideas of how to do it differed, but the goal remained intact and expressive generally of society's aim. Furthermore, this attitude toward education —the power for creating the ideal individual—seemed typical of the Greeks from their earliest literature through their decline, when it passed on as a formative concept to other nations.

The *Iliad* and *Odyssey* provided the basic educational text for Greek civilization. Homer's heroes displayed a magnificent code of knightly honor,[37] according to which they were responsible for their own conduct though gods might occasionally assist or deter them; they cared above all for their individual prowess and their rights as dignified, competent human beings. These attributes, according to Homer, were not necessarily gifts from the gods to the heroes, but were taught to them and developed by constant exercise of the wit and body. So Achilles' extraordinary abilities and powers were learned from his teacher Phoenix (*Iliad* 9.442). Similarly, Telemachus was taught to be "a responsible aristocrat" and the first four books of the *Odyssey* described his education.

> It is in the sense that the Homeric ideal of accomplishment and courtesy pervaded Greek education that Homer may be spoken of as the Bible of the Greeks. A character in Xenophon says (*Symposium* 3.5): "Because my father wished me to become a gentleman, he made me learn the whole of Homer, so that to this day I can recite the whole *Iliad* and *Odyssey* by

heart." Plato acknowledged that Homer had educated Greece (*Republic* 10.606e) but would exclude him from the ideal state nevertheless.[38]

Plato's ban on Homer and poets generally is understandable when one remembers he considered them a threat, or at least a nuisance, to the development of his ideal individual, who was quite different from Homer's. Although Plato's perfect man was also to achieve excellence and distinction, as the older concept required, his goal was to reach truth and virtuous behavior within a strongly authoritarian state. His aristocratic, ideal individual was responsible for his own accomplishments but he was allowed less individualism; he was to perform well his role within a beneficial government. Such a demand would have been unthinkable to Achilles some hundreds of years earlier. Thus the ideal for individuals changes with the years, the values of society, the degree of civilization attained, and the thought of various minds.

In the brilliance of the 4th century B.C., when Plato taught in his Academy on the outskirts of Athens, Isocrates also established his school along somewhat different lines, and in the next generation Aristotle, who had studied 20 years in Plato's school, set up his own Lyceum to promulgate his ideals. Shortly thereafter Zeno (335–263 B.C.) began his famous teaching in the *Stoa Poikile*. Wherever learned men assembled, teaching occurred in the exchange of opinions. When young boys were sent to school, they received a carefully designed education, planned to inculcate morals and an appreciation of virtuous behavior as well as develop their bodies and mental agility. Before specialization set in under the sophists' influence, the educational goal generally was a man of many accomplishments, of culture; he was not taught only a narrow skill with which he was expected to be content for the duration of his life.

Prominent among the means recommended and employed to effect this broad educational goal was the study of illustrious men: "The encomia of ancient famous men, which he [the student] is required to learn by heart, in order that he may imitate or emulate them and desire to become like them." So Plato counseled in the *Protagoras* (325d), and such materials were already available for the student and public. Historians like Herodotus and Thucydides had not recorded great events and names of outstanding leaders without character description and moral comment; and Xenophon, Plato's contemporary, presented his famous *Cyropaedia* (*The Education of Cyrus*)[39] in the same century. Although Xenophon's account becomes slightly monotonous in eight books describing the elder Cyrus' behavior as commander of a great army, the dramatic portrayal of his dynamic personality compels the reader to consider his every act: his military strategy, psychological handling of his enemies and defeated foes, consistent justice and fairness in all his relationships, his humor and clever conversation at the dinner table, and his manly restraint and respect in protecting the beautiful Panthea, wife of the Assyrian Prince Abradates. Basically the *Cyropaedia* is fiction with occasional historical underpinnings and a strongly didactic purpose. Within the story there are certainly an ample number of exemplary actions for the schoolboy to recall and from which he could safely pattern his behavior. Xenophon intended his work for exactly this purpose: he wished to depict the ideal monarch, the admirable leader who could weld widely disparate peoples into one effective empire partly through fear but also because of their "lively desire to please him." Believing such a man to be deserving of highest esteem, Xenophon investigated "his origin, what natural endowments he possessed, and what sort of education he had enjoyed, that he so greatly excelled in governing men."[40]

The story of Cyrus begins with an analysis of these points. His education, we are told, is to follow the rigorous Persian system. From early youth on, boys are disciplined in the arts of self-control and moderation, the techniques of warfare and handling of the bow, spear, and other like weapons. In the period of late adolescence they accompany the monarch on hunting expeditions and learn to endure extreme heat and cold, hunger, and physical fatigue. It is a rigorous training and though justice is prominent as a principle to be learned and practiced, the educational program is primarily military and intended to instill the sense of duty and responsibility to state authorities. This, of course, was no Persian program; neither was the constitution Persian as outlined by Xenophon. Both were strongly Spartan, the Lacedaemonian system was simply assigned to the Persians by Xenophon.

The ideal was his concern, not the fact; perhaps, as Aulus Gellius claimed, he desired to counter Plato's *Republic* with an ideal state of his own making.[41] To be sure, Socratic influence held in the wise teacher Tigranes and was seen in Cyrus himself as he thoughtfully analyzed his problems; also the Spartan model which appealed to Plato was obviously here, but Xenophon's state was created as well from his personal experience in war, and military organization became pre-eminent. His state was less concerned with abstract discussion of truth and the philosophic acumen of a council of elders. It was rather a story of the ideal in action to prove its success, its power in performance even as Socrates had asked for evidence that the state outlined in the *Republic* could be strong in times of trouble.

Thus, although the title indicates a work devoted to education this, in fact, is only a small portion confined to book one; the remaining seven books cover the full life and experience of Cyrus until his death. It is a romantic biography, the earliest of its type. The hero, Cyrus—the ideal individual—who was heralded in story and song was "most handsome in person, most generous of heart, most devoted to learning, and most ambitious, so that he endured all sorts of labour and faced all sorts of danger for the sake of praise."[42] He was a thoroughly Greek hero regardless of Persian and Median parentage, and the story of his great virtue has endured as a favorite among tales of famous men. Xenophon was legitimate reading in the Renaissance, even for ladies whose book lists were carefully censored by their guardian-teachers. His works were admired for their "sweetness" of style, their moral philosophy and good sense, and the *Cyropaedia* particularly was chosen as a "pattern for the best Prince."[43] So it was recommended reading for the young King Edward VI of England; the royal tutor Roger Ascham read it with Elizabeth, and James I added his praises and requested Philemon Holland to translate it into English for the young Prince Henry.[44] Xenophon, though less popular than Plato, Thucydides, and others, had framed an ideal portrait that fitted well into the Renaissance collection of books of manners and courtesy, stories of famous knights and sophisticated courtiers, heroes of military expeditions and Mirrors for Magistrates—all intended as examples for the young to follow. This type of instructional literature was considered vital, at least through the 19th century, to the process of promoting high ideals in the next generation.

And the Greeks made many contributions to the immense collection of exemplary literature. Cyrus' achievement pales when one considers the incomparable legends of Alexander the Great and his lasting glory. Although Sir Thomas Elyot, noted educator and courtier to Henry VIII, suggested in his book *The Governour* (1531) that Xenophon's life of Cyrus be compared with Quintus Curtius' life of Alexander the Great to evaluate the relative virtues of the two noble princes, it would have been a merely pedantic exercise because Alexander's fame so far outshone the Persian's. Cyrus, however, along with Achilles, Heracles, and Semiramis, was one of Alexander's heroes, appealing to the romantic side of his personality, which fascinated contemporary biographers as well as Napoleon and all readers since. Perhaps Alexander truly felt himself chosen by the gods; in one instance, he even asked for deification from his conquered peoples, but it was a strategic move to gain acceptance for an unpopular decision he had just made.[45] The intriguing combination in his character was that he was a dreamer, yet an extremely realistic, energetic, and effective commander. W. W. Tarn, whose thorough studies of Alexander have helped to separate fact from myth in his life, has put it thus: "To be mystical and intensely practical, to dream greatly and to do greatly, is not given to many men; it is this combination which gives Alexander his place apart in history."[46] Whether this rare synthesis of two seemingly opposite qualities accounted for his stupendous achievement no one can know, but his actual accomplishments speak for themselves: Alexander set up the structure for unification and exchange—maybe on Persian roads—but communications were established between East and West in the uniting force of his personality. The groundwork was laid for Hellenization and the Roman Empire as well as Christianization. In this sense he started a new epoch as great as any revolution since. Greek civilization and science, which he fostered (even giving research funds to his former teacher Aristotle),[47] were enabled to spread unbelievably and gain thousands of adherents.

Jewish lore made Alexander a precursor of the Messiah. And it was in this spirit he prayed at Opis that "Macedonians and Persians might be partners in the commonwealth and that the peoples of his world might live in harmony and unity of heart and mind."[48] Although the prophet Amos had implied as much when, nearly 500 years earlier, he claimed that God's concern extended to all peoples, Alexander implemented it and made it feasible. As Professor Tarn put it:

> He found the Ideal State of Aristotle, and substituted the Ideal State of Zeno. It was not merely that he overthrew the narrow restraints of the former, and, in place of limiting men by their opportunity, created opportunities adequate for men in a world where none need be a pauper and restrictions on population were meaningless. Aristotle's State had still cared nothing for humanity outside its own borders; the stranger must still be a serf or an enemy. Alexander changed all that. . . . Perhaps he gave no thought to the slave world—we do not know; but he, first of all men, was ready to transcend national differences. . . . And the impulse of this mighty revelation was continued by men who did give some thought to the slave world. . . . Before Alexander, men's dreams of the ideal state had still been based on class-rule and slavery; but after him comes Iambulus' great Sun-State, founded on brotherhood and the dignity of free labour. Above all, Alexander inspired Zeno's vision of a world in which all men should be members one of another, citizens of one state without distinction of race or institutions, subject only to and in harmony with the Common Law immanent in the Universe, and united in one social life not by compulsion but only by their own willing consent, or (as he put it) by Love.[49]

There can be no doubt that Alexander himself and his influence were two of the great forces anticipating the period of Hellenization and subsequent united, cooperative movements in the Mediterranean world and beyond.

Shortly we shall return to the interpretation of Aristotle's and Zeno's doctrines, but, at present, the question is posed: what is the place of heroic figures like Alexander and Cyrus in the utopian tradition? Tarn directly involved Alexander in the building of an ideal state, and, of course, the numerous legends of his life immediately joined the corpus of exemplary literature in which Cyrus was already established. The literary form of biography, often embellished for effect, developed rapidly in this period, partly because the superhuman Alexander stimulated it. But earlier, Isocrates and his school of thought had encouraged "the glorification of important personages," and the writing of lives as a study of character had been promoted by the peripatetic tradition (such as Theophrastus' *Characters*), which finally culminated in Plutarch and his ever popular, instructive *Lives*. It seems an oversight to bypass Plutarch's lives of Lycurgus and Solon, major contributions to this strain of thought; but the examples cited are sufficient to introduce the ideal man into the stream of thought that formed the heritage for utopia.

The ideal individual may be the goal toward which the society strives or he may be the man wise and powerful enough to establish the perfect society; in either case, the analysis of factors operating to make an ideal individual leads naturally to the consideration of what makes an ideal society. When the heroic man is encountered in literature as a forceful entity, he is evaluated not only as to his influence on affairs, but also as to his origin: what made him, what forces combined to fashion the perfect individual. Given natural endowments of health and body, education was often the formative force. Major utopias will not neglect this method of making people sufficiently admirable to carry on or live within the ideal society. Yet the reader will observe that education has received little attention until now in this study. Not one of the various utopian visions examined thus far has considered carefully the education of its citizens. Ideal societies have been a *fait accompli;* no method has been necessary. Iambulus encouraged learning generally, even as he provided abundance for physical needs so that leisure was available for the pursuit of studies; so also Plutarch explained that Ogygians had productive sufficiency and hence time to follow the demands of science; but in no instance has an educational program been devised to implement the utopian society, to train youth to carry it on, much less to create it. Thus it was from the classic stories of ideal individuals, which included analyses of the forces shaping a perfect man, that education received emphasis and, from this idealistic stream of literature, entered the realm of utopia.

The Ideal Utopia

Although the subtitle may read like a double affirmative, it is in fact an accurate designation for Plato's *Republic*. Not only was this the first full-scale utopia and the archetype for future writers like Bacon, Hartlib, Milton, and others who defer to it, but it was purely "ideal" with no hope or thought expressed for its establishment on earth. There was no question of claiming its existence on an unknown island at the world's end, no attempt to create a false reality or authenticity for the work. (This characteristic of later utopias came, as we have observed, from other sources.) Instead, the new city was frankly placed in the philosophical realm, and its structure arose from hypothetical discussion guided by Socrates' discerning mind, so that each element of the society was debated and justified before inclusion. When the city had finally been conceptually constructed, Adimantus, one of the participants, commented that it was a city whose home was in the ideal, for "I think that it can be found nowhere on earth." And Socrates replied, "Perhaps there is a pattern of it laid up in heaven for him who wishes to contemplate it. But it makes no difference whether it exists now or ever will come into being." A man can order his life by its laws.[50] The *Republic* thus was a transcendental pattern which, once understood, could serve as a guide to man. The aim was primarily ethical. Moreover, the theoretical nature of the city was great enough to cause Socrates on the following day to ask that the image be put into motion, that the ideal be animated to prove its effectiveness in action, which resulted in the dialogues recorded in the *Timaeus* and *Critias*.

As an ideal society, the *Republic* has the further distinction of unifying many of the various strands, both religious and secular, forming the utopian heritage. Here all converge and are woven into a meaningful pattern: the *Republic* pictures an ideal people living a natural life of simplicity unspoiled by luxury or ease; Socrates is certainly the ideal individual; and the whole description concludes with Er's vision of the rewards for a just man in the heavens beyond. It is also the most thorough coverage yet given to an imaginary society, and education plays an important part. The content is too well known to require detailed review. The communal pattern of living without marriage and private loyalties to the family was seen reflected in Iambulus' Sun State; the organization of social castes was adopted by Euhemerus but the categories were changed. Rather than rule by the priestly group, Plato set up the counselors or guardians of the state who supervised activities of the military and artisan classes. It was an absolutely aristocratic society with leadership based on intelligence and character, not nobility of blood or inheritance. And the educational system was to prepare the selected few for their duty to the state. The Spartan model behind Plato's well-controlled, disciplined state was not hidden, but displayed as the strong pattern necessary in a period of disorder such as Athens was then actually experiencing. No doubt Plato's distress over the ruin of Athens' greatness influenced his creation of a political utopia, for it is interlarded with comments not always oblique on the present state of affairs in his own home town. Moreover, the Greek city-state became the basic unit for the *Republic*, and the general climatic and geographic background of his native land was assumed. For example, the participants in the dialogue realized the practical impossibility of self-sufficiency for their ideal society, recognized the need for imports and trade, and hence allowed shopkeepers and money exchange to continue in the agora or marketplace. But gold and silver were not to be touched by the state's guardians, nor were they to own possessions beyond their need. This was to prevent temptation and personal greed, thereby helping to insure their constant devotion to the public good and "the greatest possible happiness of the city as a whole." Each man contributed according to his individual nature, and his work as an artisan, farmer, or tradesman was specialized, since it was impossible for one man to do many tasks well. Thus as everyone added his portion of service, the total state attained harmony and proportion, even as a statue in which each part rendered its own function and the whole became beautiful.

To achieve this state of happiness, the best individuals must be carefully nurtured and properly educated. Unless a man was made perfect, the state could never be ideal. The school system then merited highest attention in the dialogue: it determined the quality of whatever followed and so was basic to the entire way of life proposed. Absolute good was the goal, and those wise old men who had learned the use of reason, the highest faculty, were rewarded with positions of leadership in society. Only the philosopher was fitted to be king, and the entire educational program was designed to improve man's capacity for higher thought.

In this sense mathematics entered the curriculum, to develop the powers of logic, to compel the soul to reason and have an elevating effect. Knowledge of the eternal, of philosophic truth, was "the real object of the whole science." Thus, although Plato's emphasis on the study of mathematics has been taken as evidence of scientific interest, since mathematics has subsequently been so important in the development of science, he had no such notion in mind, at least not in the *Republic*. There was no suggestion that the powers of abstraction inherent in mathematics should be used for unraveling the laws of the universe regardless of the fact that his neighbors in Greece—the Milesians—were already applying it in this very way, and Plato himself employed mathematics extensively when explaining the universe in the *Timaeus*. This will be considered later (Chapters IV and

VII), when scientific knowledge reached a point of formative impact on society. In the *Republic* and other utopian visions of Plato, scientific interests, the claims of natural philosophy, and the investigation of natural phenomena, have not invaded the culture. The same author may elsewhere significantly advance the work of science, but he has seen fit neither to incorporate it directly into his educational system nor to provide for its advancement in the utopian society at large.

This statement, however, requires qualification because Plato, first among dreamers of utopia, provided a method for the progression of knowledge—the dialectic method—by which he built the *Republic* itself. It was a logical procedure of causal reasoning starting from the particular and moving up the hierarchical path to truth, sometimes descending to corroborate the findings. Immediately, mere opinion was ruled out as subjective and irresponsible; in its stead an orderly ladder of thought was installed, inductively leading to a higher and more inclusive principle. Plato inherited the system from Socrates, but he made it his own by going further: Socrates stopped with the particular[51] and ended the questioning process with ironic ignorance that prepared the pupil for unbiased, logical thinking, while Plato went on with the cautious investigation from particulars and combinations of them to higher truths; but he, too, often fell short of expressing the final concept because the highest truth could be seen only from a distance,[52] the beyond was unknowable. At this point his mystic, poetic side entered and perhaps prevented folly. In any case, it preserved his dialectic system from hardening into an easily refutable method, allowed numerous interpretations of his procedure, and kept the feeling of life and change, development, in his writings, which extended their immortality as civilization made discoveries and moved ahead.

Employing Plato's own method, we may better evaluate its position regarding science if we consider first the particulars: how the dialectic method operated and then where—under what circumstances—he used it. First, it was inductive, accumulating evidence step by step in definite order. This has been one effective approach in building scientific knowledge, and it is the general method Francis Bacon endorsed. But what determined the order for Plato? The dominant control lay in the leader's mental powers, that is, Socrates'. It was not a laboratory for investigation without preconceived hypotheses; on the contrary, the so-called "particular" was stated as a conditional hypothesis to be upheld or dashed upon analysis. The goal of the system was to reach the unconditional, the absolute truth and Reality, which lay in the Idea for Plato. Further, the entire guided process had to conclude with a concept amenable to a purposive or teleological universe in which we are each a part as well as a microcosmic reflection of its functions. The framework then for the dialectic method was pre-existent and determinant in the last analysis. This was not free inquiry in science regardless of its pose as unrestricted questioning. It was inductive in appearance but it was also deductive from its relationship to the great cosmic scheme of things. Now, still within the area of operation, what is the nature of the matter, the materials, or substance with which we are dealing? Plato as a philosopher was working with words, forming his synthesis from the examination of such elusive materials as behavior, ethics, and concepts. As he himself put it, he "nowhere makes use of any sensible object but proceeds wholly from ideas, through ideas, to ideas."[53] This was to him the proper approach to philosophic truth, and the dialectic faculty was well designed for this purpose. Thus it proved to be both a powerful teaching method when used by Socrates and an effective means of constructing a system of unified thought when utilized by Plato as in the *Republic*.

Secondly, it is instructive to observe where Plato used the method in order to see its relationship to scientific inquiry. Here the evidence is most definitive, for, in proportion as his subject matter becomes more scientific, he employs the dialectic method less, preferring straight exposition, a lecture. This is precisely what occurs in the *Timaeus*, the most scientific of his writings. When the principles of the universe and its creation are to be set forth, Timaeus simply discourses without interruption of questions or acceptance of the argument by his listeners. Mathematics is the method for explaining the universal, and the dialectic method is withdrawn.

The differentiation in methods according to the purpose and matter being dealt with could be considered an early example of two roads to truth—mathematics for the scientific area and the dialectic method for the ethicophilosophic—but such a thought would never have occurred to Plato, and it would be an oversimplification, for he also used myth effectively to explain the unknown operations of the universe. All knowledge was unified for Plato: scientific, moral, and intellectual aspects were combined into a significant whole, a philosophy of the ideal vibrant with the progressive thought of the author and the questioning brilliance of his revered teacher Socrates.

Hence the dialectic method that created the structure of the *Republic* was neither scientific, as we think of method today, nor was it chosen by Plato when he considered the physical phenomena of the universe in the *Timaeus,* but it was important as a logical, orderly process.

Still Plato's ideal society remained most conservative and adverse to change. To be sure, the ideal was a reality: the society was led by philosophers and based on education for virtue and rational contemplation of the good. The concept of progress was operative only in the process of achieving the goal. Once perfection was reached, the situation became permanently fixed. The modern idea of progress, as we know it, had not dawned, nor could it have a place in the perfect scheme of things. Innovation was frowned upon as dangerous to the stability of the social group. Plato's Republicans believed that the spirit of license could slip unnoticed into society even through a new mode of music. Such lawlessness might spread and cause fundamental changes in the laws, eventually resulting in reckless manners and customs. So the citizens zealously protected the status quo and lived in immutable perfection.

Aristotle, of course, severely criticized Plato's *Republic,* differed with him on the question of private property and the family, and said there was too much unity in Plato's state and uniformity among citizens; hence it was impractical, undesirable, and good sense would repudiate it.[54] Partly as a corrective then to Plato's errors, he outlined in his own logical method the ideal *Politics* in which he too defined the state, its structure, categories of citizens, including the natural slave or servant class, and the educational program to train the virtuous ruler. As might be expected from Aristotle, he stressed the particular fact, and analyzed existent states and results of various forms of government. His approach was departmentalized as well and lacked the fusion, the unity of Plato's presentation; it is a dissection of the state; veins, arteries, and skeletal structure lie exposed on the operating table. It is more comparable to the form Plato used in his *Laws*, which is expository. While the ideal is discussed, there is neither the fictional element nor other characteristics of the utopian genre. These works instead start the long tradition of writing ideal constitutions which, at their closest, are cousins to the utopia and, at their most distant point, are the actual statutes of government like the Constitution of the United States. Such works, however, fed the minds of utopists when they decided upon the best state for their imaginary people.

Among the many writings of this sort left by the ancients was Zeno's *Republic*, which Professor Tarn compared with Aristotle's ideal state, pointing out the cosmopolitan attitude of Zeno and the relatively narrow, nationalistic concern of Aristotle. The contrast is well drawn and we may add that Zeno (335–263 B.C.) went considerably further in adding new concepts for utopia. As the outstanding leader of Stoic philosophy, the teacher of the Painted Porch, who competed with Aristotle's peripatetic school and Plato's successors in the Academy, Zeno presented an ideal state markedly different from the other two. While he reflected certain ideas such as Plato's community of wives, he called for the complete abolition of coinage, a thorough overhauling of the current system of education, and the end of class distinction. His was a community of all mankind and perhaps even more "ideal" than Plato's. There were no temples, law courts, or gymnasia; adornment of the state lay not in costly offerings or impressive buildings, but only in the virtue of its inhabitants. All people could progress from folly and ignorance to virtue and wisdom. And over all in this world-community presided Eros, the god of friendship and concord.[55]

Zeno certainly proposed more extreme treatment than Aristotle and Plato for the cure of mankind's ills; nevertheless, the three great thinkers were in general accord in the final analysis of their utopian views. All maintained the goal of virtue built upon knowledge and understanding. This was the greatest need of man, the answer to his search for happiness and thus the aim for an ideal society. Furthermore, though each of these philosophers contributed to the growth of scientific thought, each proposed his method of approach to knowledge and directly aided the study of certain scientific areas, still no one of them placed science in his ideal state. Ethics, politics, and education were both the means and the end for utopia.

From secular views of the ideal—with religious foundations—have come the elementary patterns and content that utopian authors have reflected for centuries. The Greeks provided a rich variety of suggestions: the ideal of nature and the simple shepherd's life as pictured by pastoral poets or the early novelist Longus, the ideal people described by historians like Herodotus and Euhemerus, the ideal individual personified by biographers in the lawgivers Solon and Lycurgus, the great military leaders Cyrus and Alexander, and finally the philosophers' truly ideal utopia first pictured by Plato. Such writings by major authors attest to the power of the ideal in their society; the paradigm, model of beauty, even though manifestly inaccessible, was constantly held before a people who worked toward excellence and perfection and realized that the responsibility was theirs alone. This body of material forms the background for all idealists including those who chose to write utopias or describe imaginatively perfect societies.

Besides the general idealistic attitude, many specifications for content and structure were bequeathed by the classical and Hellenistic writers to utopists to come. On the surface, there were geographic locations for ideal lands, clever techniques to simulate reality and give the reassurance of authenticity, city-plans or architectural layouts available for future designers, programs to promote health, longevity, and desirable offspring, and the ever-prevailing abundance and balmy climate conducive to the happy life. Below this level, the conceptual bases are much more complicated. In the subterranean depths of the history of ideas we can point out only those images that stand forth quite clearly in the light of progress. Science was not yet among them; such investigations had neither sufficiently permeated society nor given enough benefits to deserve a position in the ideal world. Moral behavior and metaphysical truth far surpassed knowledge of the physical world as a means for man's betterment. An attitude, however—the belief in a logical approach to increase learning—has entered man's consciousness as an educational means and as a way of extending absolute knowledge. This is of utmost importance, because it not only endorses the religious foundations for a rational, comprehensible order of things, but it encourages man's attempt to understand the order. Once this principle is accepted, new means will be found to decipher the mysterious universe, many codes or methods may be invented and each may have its effectiveness. And the ideal world of utopia too will change with man's discovery of new reality.

III

THE RENAISSANCE AND TRANSITION

> Now, let Italian, and Latin it self, Spanishe, French, Douch, and Englishe bring forth their lerning, and recite their Authors, *Cicero* onelie excepted, and one or two moe in Latin, they be all patched cloutes and ragges, in comparison of faire wouen broade clothes. And trewelie, if there be any good in them, it is either lerned, borowed, or stolne, from some one of those worthie wittes of *Athens*.[1]

So ROGER ASCHAM, that remarkable man of learning who had read Xenophon with Queen Elizabeth, expressed the opinion of many humanists in the 16th century. And not only did he believe it, he fantastically claimed that Athens contained more learned men in a short period "than all tyme doth remember, than all place doth affourde, than all other tonges do conteine." Perhaps Ascham overstated the case, but there was no question that the Greeks carried a heavy weight in authority and provided an almost indisputable challenge to all but the hardiest thinkers, the most presumptuous of the "moderns" in that time. Renaissance scholars were exceedingly impressed by the Greek works they read and so placed the classics high as models to be imitated and vital to the educated gentleman. And from the Greek models, came an extremely adequate provision of materials for utopias.

Renaissance utopists could not draw so effectively upon Roman letters, rich as they were in maxims for guidance and idealism generally. The Romans did not visualize utopia. Their genius was in practical methods of organizing, communicating, extending jurisdiction, formulating laws and civil controls. Vergil expressed the Roman attitude when he said, "Let others mold the breathing bronze, plead causes and tell the motions of the stars. *Your* task, O Romans, shall be to govern nations, to spare the conquered and defeat the proud." Adhering to this belief, they built a great empire on the groundwork laid by Alexander and the Greek standard of perfection, which they readily adopted. The Romans produced many competent, realistic men of affairs who understood ways and means of creating a cohesive, powerful civilization; but they bred few dreamers. One feels they were always too active, too energetic to dream; they seldom expressed the desire for leisure to pursue philosophic studies as the Greeks did. Instead, expediency was foremost.

This attitude clearly affected scientific endeavor as well as the creation of utopias. What the Romans added to scientific knowledge compared to their Greek and Alexandrian predecessors was practical and technological, not theories or abstract ideas that might have promoted new thought. Their advances were in surveying, map-making, reckoning the calendar, sundials, and solutions to mechanical problems of an immediately useful sort. Even in these areas the Romans were most dependent on work of the Greeks who continued to dominate their conquerors in intellectual thought and in language as well for many years. In more theoretical fields like astronomy, mathematics, geography, medicine, and botany, the progress resulted from works by men of Greek or partial Greek descent: Hipparchus and Ptolemy, Strabo, Galen, and Dioscorides in his five books on plants, *De Materia Medica*.[2]

During the period extending roughly from the 2nd century B.C. to the fall of the western Roman Empire in the 5th century A.D., the Romans were compiling, gathering together borrowed materials into impressive encyclopedias, and developing a highly artistic style of presentation in their Latin language. In one sense the collected volumes were most valuable in preserving past knowledge, but in another way they rendered a disservice to scientific advancement. Lucretius' *De Rerum Natura*, the collection of the elder Pliny in his 37 books on *Natural History*, other encyclopedias by Varro and Vitruvius, and Celsus' *On Medicine*

—all continued as basic books in the curriculum throughout the Renaissance, thereby offering authority in their erudition and extensive coverage. But they provided no great incentive for others to verify, to observe, to find proof; they gave the facts as answers embedded in a strong philosophic setting. The Latin authors' interest was not in evidence, method, or mathematics. Nevertheless, some information from the Greek period was saved, and the powerful gleam of Roman reason and order emanated from these works which, in itself, was a gift to the rational thought needed in the composite environment from which science emerged.

The scientific historian Charles Singer attributes the Romans' general "failure in inspiration" to the Stoic and Epicurean philosophies, which were strong among their intelletcual leaders.[3] Neither school of thought was especially encouraging to new scientific ideas, and both saw man's life controlled by external forces, whether these were the spheres and astrology, as Stoics believed, or the more materialistic determinism of the Epicurean doctrine based on the atomic structure of the universe. Both were determinate and left man in the position of accepting his lot (either with moral duty and strength of will or with hedonistic ethics), not attempting to change it. Man's acceptance, however, did not always preclude his curiosity and effort to find out or explain the effects of the outside force. An outstanding voice in opposition to the finite concept was Seneca's, when, in his *Natural Questions,* he enthusiastically expressed his belief in the progress of knowledge and the unknowns to be conquered by the intellect:

> We enter a temple with all due gravity, we lower our eyes, draw up our toga, and assume every token of modesty when we approach the sacrifice. How much more is all this due when we discuss the heavenly bodies, the stars, the nature of the gods. . . . And how many bodies besides revolve in secret, never dawning upon human eyes? . . . Many things, moreover, akin to highest duty or holding power near it, are still obscure. . . . [Yet] how many animals we have come to know for the first time in our own days! Many, too, that are un-

known to us, the people of a coming day will know. Many discoveries are reserved for the ages still to be, when our memory will have perished. The world is a poor affair if it do not contain matter for investigation for the whole world in every age.[4]

While this sounds uncannily like some statements to come centuries later when men were advocating investigation of nature's phenomena, Seneca was content with the mere expression of the thought, and his actual work contributed little to new knowledge. His sources were borrowed, and his total view was purposive and ethical in emphasis. Though he criticized Stoic belief in some aspects, he remained true to it in others, especially in its moral interpretations. Thus, in spite of the appearance of belief in the advancement of learning and faith in human destiny, Seneca did nothing "to stay the downfall of ancient science."[5]

Following the Epicureans, Lucretius made the greatest contribution: his firm faith in reason and nature's performance according to natural law. This the Western world inherited with due respect. Other less important examples can, of course, be found in Roman literature showing advanced thinking, such as Cicero's argument against superstition and Celsus' pointing out the valuable distinction Hippocrates had made in separating medicine from philosophy. But, generally speaking, the Romans were no more creative in the scientific world than they were in the creation of utopian worlds. Undoubtedly, their deep belief or faith in the grand destiny of the Roman race also made fanciful visions of better worlds completely useless. Aeneas had behaved in exemplary manner; he practiced self-denial, accepted fully his social responsibility and courageously founded the nation according to plan. Hence their aim was not only clearly defined and promised but it was widely accepted. When people know the direction in which their society is moving and feel sure it is best, there is hardly any reason to create dream worlds or to search for new knowledge unless it solves the practical problems encountered en route to the accomplishment.

The same potent factor of belief—directed differently—operated in the Middle Ages and, similarly, few utopias were written, and major advances, earth-shaking discoveries, were not made in science. A completely structured universe, divine in purpose, dominated the whole fabric of thought and, again, man had a most definite position: he was to follow set rules for behavior that allowed little leeway if he would avoid hell's terrifying fires and qualify for heaven's reward after his precarious life. Man's attention, therefore, centered on God's will and purpose.

Of course idealism was prevalent in codes of chivalry and in saints' legends, which gave examples for imitation and established goals, but the only major utopias (if we extend the definition of the genre) presented in the medieval period were St. Augustine's *City of God* and maybe Dante's "Paradiso" in the *Commedia*, if we consider him as a final spokesman for the Middle Ages rather than the introducer of the Renaissance. In either case, his vision of heavenly paradise and St. Augustine's philosophic view of the ideal city were both aimed at defining man's relationship to God and His universe. Both were religiously oriented and highly mystical; both described how the individual could achieve unity with God. Although Dante explained that his work was "undertaken not for speculation but for practice," the purpose, as he also stated it, was "to remove those living in this life from a state of misery and to guide them to a state of happiness."[6] The "Paradiso" remains a monument to faith in God, the power of Love, and an ethical chart to guide man. Comparatively, Augustine's *City* had the same purpose and cast all hope in God.

This world, limited and subject to destruction, was scarcely worth knowing, even less worth examining carefully or systematically for new ideas. This is not to suggest that nothing occurred toward scientific advancement, that the long period between the fall of Rome and the Renaissance was truly all Dark Ages or Gothic fog.[7] To the contrary, many contributions were made, many questions asked about natural causes, and investigations carried on. Knowledge new to the West, as well as translations and compilations of Greek writings, was entering especially from Islamic sources. The School at Chartres, which studied Plato's *Timaeus* with emphasis on its mathematical and systematic explanation of the universe, also welcomed Aristotle's and Ptolemy's works. By the middle of the 13th century nearly all important works of Greek science were available in Latin translation. With all this, however, the scientific development of the period seems to have generally continued the Roman tradition with focus on practical interests, compilations, and the revival of the ancients' writing. Important as this was, and there were "modern" men like Roger Bacon who saw the nature of scientific method and the power inherent in it, still the work done did not result in highly original discovery or in philosophic foundations that could take the test of new physical facts. While for some individuals there is evidence of a turning toward material objects, away from a moral, philosophic outlook, it was, nevertheless, the latter view that held for the great majority. And it was Aristotle's thought that was chosen especially for ecclesiastical purposes. Aquinas most effectively did the job, and Aristotle's structure for the universe was firmly installed to support the medieval heaven that continued to be man's heaven until the 17th century, when it came tumbling down with loud noises as well as low, reverberating rumbles in the land of intellect. It was an exceedingly powerful structure, clearly outlined and satisfying in its all-inclusive combination of man's beliefs—all snugly fitted into the perfect circle of design. It was a teleological universe, each part had its directive purpose; the whole was a living entity, and the microcosmic man could easily see himself reflected in analogous parts and functions of the macrocosmic universe. It was to be expected that the greater, purer heavenly bodies should influence, if not completely control, the lesser life of man. So the study of astrology, which was closely linked to astronomy during this time, was encouraged. On this great scheme the scholastic thinker could elaborate *ad infinitum*, and he did. Rhetorical logic flowed freely and magnificently in completing and adding to it. Often it seems the synthesis was more important than its parts; the whole was sought more than veracity or accuracy in the building process. In this development Aristotle, as he was understood generally through his interpreters, was irrevocably joined to the scholastics' approach.

No one explained the resultant universal structure better or more beautifully than Dante in his *Divine Comedy*, built graphically from his descent to the center of the earth on up through ever-ascending circles past the crystalline sphere of the primum mobile to the final empyrean, heavenly paradise. As Dante presented the concept of the medieval universe, so he also expressed its spiritual purpose. The utopia was once more a heavenly paradise, not oriented secularly or proposed as possible on this earth, though some ethical lessons might have been learned. Since it was only above the firmament, it offered few building materials to the utopists of the future. For these the Renaissance author—like the Renaissance scientist—turned directly back to the Greek classics, and Plato, for the utopists, was recognized to be the great master of ancient form.

Christian Virtue

Among early readers and admirers of Greek literature in England was Sir Thomas More whose Utopians, descended from Greek stock like their ancestors in Plato's *Republic*, lived in an ideal philosophic state in which the pursuit of virtue was the main goal:

> They dispute of the good qualities of the soul, of the body, and of fortune, and whether the name of goodness may be applied to all these or only to the endowments and gifts of the soul. They reason of virtue and pleasure; but the chief and principal question is in what thing, be it one or more, the felicity of man consisteth.[8]

Although More's Utopians continued Plato's interest in the rational consideration of ethical questions and were completely convinced that man's happiness came from virtuous living, the society of *Utopia* was very different in several respects from the Platonic paradise conceived nearly 2000 years earlier.

One of the changes was More's genuine concern for the solution of practical, economic problems,[9] which reflected the growing secularization of thought in the Renaissance. Poverty and taxation, money and the apportionment of wealth, were basic issues requiring answers in *Utopia.* More's account, in contrast to the idealistic *Republic,* which had little time or space for business matters in the agora, devoted much more attention to the problems of daily living. His planning stressed man's industry and hard work, not merely his right thinking. His was not a land flowing with milk and honey; men must work and plan to provide the plenty they wished to enjoy.

Innovation was no longer discouraged as in the *Republic,* but still it was not especially rewarded or singled out for attention. To be sure, the Utopians had discovered "a marvellous policy" by which hens were relieved of the job of sitting on their eggs; instead, eggs were automatically hatched by artificial heat. The citizens had also devised instruments that *exactly* measured the movements of the stars and planets, and they had invented remarkable engines for war that must be kept secret until needed in military action. Though More cites these three specific inventions, and the people of *Utopia* were quick to learn the production of paper and the art of printing, no statues were erected to inventors. In the marketplace of the capital city, Amaurote, statues were of notable men who had been benefactors of the commonwealth. The images were simply "for the perpetual memory of their good acts, and also that the glory and renown of the ancestors may stir and provoke their posterity to virtue." They were in the tradition of ideal individuals seen earlier.

Sir Thomas More also had a complicated philosophic issue that Plato of course did not have to face: the basic tenets of Christianity. Here, in opposition to the ascetic and negativistic teaching of some medieval churchmen, More asserted the Renaissance humanist position and argued for pleasure and joy in life. God in His bountiful goodness intended happiness for man on earth; the joyous life was not withheld for heaven alone; it could be reached through the practice of virtue, which was a life ordered according to nature's reasonable rules. Man was intended not to inflict torture on himself to atone for his sinfulness, but to enjoy the good and honest pleasure of a life that did no harm to others. Tolerance prevailed even in religious matters: Utopians might hold different beliefs, each following his own reasoning, and anyone who became too zealous in affirming his position and condemning that of others was promptly exiled. Only gradually after the visitors introduced Christianity were they beginning to believe in Christ and the word of God. And most important, Utopians were diametrically opposed to the type of attitude Enoch had expressed in pre-Christian days, which had endured and gained a majority of supporters, namely, that man should not presume to understand God's laws, that such an attempt would repeat the audacious act of Adam and Eve in their fatal disobedience. On the contrary, Utopians wholeheartedly believed that God approves of human curiosity:

> For whiles they by the help of this philosophy search out the secret mysteries of nature, they think themselves to receive thereby not only wonderful great pleasure, but also to obtain great thanks and favour of the Author and Maker thereof, Whom they think, according to the fashion of other artificers, to have set forth the marvellous and gorgeous frame of the world for man with great affection intentively to behold. . . . And therefore He beareth (say they) more goodwill and love to the curious and diligent beholder and viewer of His work and marveller at the same than He doth to him which, like a very brute beast without wit and reason, or as one without sense of moving, hath no regard to so great and so wonderful a spectacle. The wits, therefore, of the Utopians, inured and exercised in learning be marvellous quick in the invention of feats helping anything to the advantage and wealth of life.[10]

On the basis of this key passage and the general attitude that More's Utopians effected social progress through their own efforts, Robert P. Adams stakes his claim that the *Utopia* advanced essentially the same argument as did Bacon's *New Atlantis* more than a century later.[11] Yet he grants that there are vital differences between the two works. Bacon's lust for knowledge, for perpetual new inventions, and constant material improvements is not found in More's *Utopia*. More's society did not in any way emphasize the investigation of nature, nor did it suggest any particular method for the advancement of knowledge. The

Utopians merely reconciled their interest in nature with God's purpose. More was, first and foremost, a Christian humanist who intelligently combined Platonic thought with the Christian tradition. His Utopia, like the Republic, was a philosophic land dedicated to the highest ethical ideal, a land which, as More himself admitted, one "may rather wish for than hope for."

Although conceived primarily in a Platonic spirit and reflective of similar goals, More's imaginative "no place" introduced certain characteristics of the modern utopia: increased attention to practical problems and the significance of man's diligence; the legitimacy of innovation; and, above all, the idea that God applauds man's use of reason in studying nature. Furthermore, Utopians lived in an evolutionary society that changed for the better, not a stationary state retaining its pristine excellence. Even as Christianity had recently been introduced and was progressively gaining adherents, so in their housing too they had advanced from living in low, homely cottages like a shepherd's to their present stately dwellings of three stories. This was a society with a history and a future. Utopians then were definitely on the road to the new society that appeared a century later in Andreae's *Christianopolis*, Campanella's *City of the Sun*, and Bacon's *New Atlantis*—though not quite so far along the road as Mr. Adams would have us believe. Their progress was still toward an ideal based on the ancient archetype, which had its staunch defenders for the next 150 years regardless of revolutionary changes in society.

The great and eventually startling advances in scientific knowledge and the possibilities it offered for man's betterment were slow in building, slow in reaching even the learned men, much less the general public. Naturally then, the new ideas emerging did not immediately displace all other human visions of the happier state. Traditional notions are hard to abandon and, during this period of transition, some utopian writers clung tenaciously to the old values and saw improvement through the pursuit of virtue alone; they either deliberately omitted mention of scientific discovery or were ignorant of the new learning.

Sir John Eliot (1592–1632) was a noteworthy example of

the group paying no heed to the world-shaking discoveries made in his own lifetime in the early 17th century. He completely ignored the new outlook in *The Monarchie of Man*, which he wrote in 1631, in fact while imprisoned in the Tower of London because of his political activities in behalf of parliamentary rights against the king's misuse of authority. Even though he was so actively engaged in current affairs of state and in this area, at least, up-to-date and courageous, he remained outside the vanguard of the new learning. When he declared his ideal in a utopian tract—a didactic essay lacking the fictional framework usual to the utopian genre—virtue alone was the prime attribute of the prince; and, in man, the microcosm of the state, rational virtue too must rule. Justice, prudence, temperance, and fortitude were the four cardinal virtues as Plato had named them, and Eliot drew typical analogies to the four streams of paradise and the four elements of nature (water, air, earth, and fire), which must be in proper combination and interaction for man to attain the *summum bonum.* He leaned heavily upon classical writers to support his doctrine.

In *The Monarchie of Man*, nature and its operations appear merely as poetic metaphor. Under divine providence the virtues in man or state are like the elements that work by sympathy and contrariety to one end—the harmony of the spheres and the glory of the planets. "Their motions are all various and different . . . but all most regularly exact . . . the secrets of whose reason what man can comprehend?"[12] Man's felicity is to be found not in the search for laws of the universe but rather in his just rule of the state within him, and the application of comparable principles of justice to the macrocosm, his nation. So Eliot, writing several years after Campanella, Bacon, and Andreae had presented their new societies with scientific emphasis, chose to follow the established and generally Platonic path to virtue. He was even less advanced than More, who had encouraged inquiry into the physical universe that could only result in greater praise for the Lord of creation. Eliot felt the secrets to be incomprehensible but the display of universal order worthy of admiration.

Other utopists of the same period were clearly aware of scientific discoveries and decided to repudiate the new ideas. Christian virtue must be maintained against the enthusiasm developing for new-fangled notions such as the search into nature's secrets. What better way than satire to show the ridiculousness of such endeavors? This is Joseph Hall's method in his *Mundus Alter et Idem* (c. 1605), which was freely translated by John Healey as *The Discovery of a New World.* Parodying English vices and customs in rough, Rabelaisian satire with great verbal relish, Hall discovers the land of Tenter-belly with its two provinces, Gluttonia and Quaffonia, and its neighboring states Shee-Landt or Womandecoia, Fooliana, and Theevingen. In Fooliana live the Foolosophers who shave their crowns so that hair will not intrude betwixt their brains and heaven. They spend their time framing fictions and believing them. Here in the valley Capritchious is located the college of Gew-gawiasters,

> who give them selves wholy to the invention of novelties, in games, buildings, garments, and governments. Hee that can devise a new game or a new fashion, according to his invention hath a place of dignity assigned him by the Duke.
>
> He that first devised to blow out bubbles of sope and spittle forth of the walnut shell, is of as great renowne amongst them, as ever was the first Printer, or Gun-founder amongst us of *Europe:* these *Gew-gawiasters,* are in great esteeme in Court, yea and amongst the meaner sort also, in so much that many of them will not put on a tatter, nor once move, without their directions.[13]

Paracelsus (known for his new, though questionable, approach to medicine and chemistry in the early 16th century) is provost of the college and credited with the invention of the strange language—the Supermonicall tongue. Under the college's pervasive influence, the people of Fooliana try all sorts of interesting experiments, such as walking down the street on their heads or hands and flying about with wings made of wax and feathers. They are fascinated by strange engines and so intrigued with the secret art of alchemy that they send to an oracle for guidance in their work. The answer given is *Travaillez,* but they fail to understand the advice, rush home as if they had their god in a box, and continue their conjunctions and fermentations until all ends in putrefaction. The poor benighted people are unable to comprehend the word which the translator renders for the reader: "Take pains."

> A mattock and a spade will get you gold
> Sooner than Chymistry, a thousand fold.[14]

Though Hall sensibly rejects the magic of alchemy, he is really ridiculing all scientific efforts, including general research on natural phenomena.

> The people hold themselves wonderful wise, and profess the search of nature's most abstruse effects; never leaving til they have drawn one reason or other, from the very depth of investigation. *They have but one eye apiece.*[15]

The citizens of Fooliana or Moronia were simply misled by the fellows of Gew-gawiaster College and wasted their time in stupid endeavors. Transposing Hall's burlesque into a constructive key, we hear him preaching a Christian life of Stoic virtue, moderation, and diligence, the same principles he advocated in his *Characters of Virtues and Vices* (1608) and earlier moral writings, which won him the epithet "our English Seneca." As Bishop of Exeter and a prominent spokesman in the religious controversy raging at the time, Hall concentrated his attention on moral and church issues; his *Heaven upon Earth* was explained in the secondary title: *Or of True Peace and Tranquility of Mind.* Although clearly aware of the excitement generated by the experimental attitude, he obviously expected from it no miraculous solution to man's dilemma; the answer lay in the individual's ethical and religious reform.

Similarly, the road to perfection was ethical behavior and Christian belief for the society of *Nova Solyma* or the New Jerusalem envisioned by Samuel Gott in 1648. These virtuous people, following traditional arguments in religion's defense, held irrevocably that God's supreme knowledge can never be understood by man. Yet they recognized the claims of the new science and feared it was getting out of hand. It was going too far, threatening God's omnipotence and causing men to become overweening in pride. Hence Gott's ideal citizens, while they did not ignore scientific studies—in fact they carried on experimental research—realized that some of God's secrets will always remain unknown. It is a tribute to the growth of experimental philosophy and its increasing importance that Gott thought it necessary to warn once more against its presumptuous claims and to guard God's throne from the onslaught of the new knowledge. He wished to "prevent art from being exorbitantly lauded as is so commonly done." So Gott stands with Bishop Hall in the conservative position even though some 40 years have passed since Hall's *Mundus* suggested the need to watch out for the enthusiasts in natural philosophy. During these years, however, great discoveries had been made, and Gott himself was directly interested in scientific matters, which undoubtedly increased his awareness of the issue and caused him to go to greater lengths than Hall in clarifying the relationship between science and religion.

Little is known about Gott's life (1613–1671) except that he was educated at Cambridge and Gray's Inn, and apparently lived in the country gentleman's fashion and served three brief periods in the Commonwealth Parliament. Most of his writings were moral and religious in purpose, but scientific knowledge is indicated in two manuscripts on which J. Max Patrick has commented. One was Gott's letter concerning speculation on the motion of heavenly bodies, which he dated October 19, 1668, and addressed to the noted mathematician and a founder of the Royal Society, Dr. John Wallis. The other was a manuscript of Wallis' entitled: "A Conjecture on Mr. Gott's Proposal for an Artificial Spring," which was to "ebb and flow every twelve hours."[16] From these papers it would seem that the author of *Nova Solyma* was perhaps one of the minor "virtuosi" of the century and had connections with at least one man prominent in the progress of science, but still he cautioned against the new belief.

Gott's mouthpiece in the ideal city is a wise and admirable

citizen, Joseph, whose task it is to educate the young foreigner, Politian, playing the role of critic and supporter of modern philosophy until he is shown the better path. Joseph speaks ecstatically with sincere belief:

> When I survey the sky and land stretching out everywhere in a huge circle (than which figure none encloses larger space); when I find no spot unoccupied, useless, or superfluous . . . the air, rarest of all elements, almost a void, giving willing way to all solids, and yet eagerly pressing into every vacant space . . . the element of fire, too, shut up like a bold, bad robber in its prison of flint, or banished to the earth's utmost limits . . . when, I say, I consider all these Divine arts of Nature, I feel to be encompassed by so many superhuman marvels that every human artifice seems maimed and cobbled; and oh! to make of *human* art an idol, how do I scorn it as an idle thing![17]

At a philosophical garden party, Joseph draws a fine distinction between "discoveries," which are merely a recognition of the nature God provided, and man's "inventions," which have been few indeed:

> Human ingenuity produces certain extras, but from no other source than Nature do they come. For what, I pray, can a cook or a physician or a chemist produce except the preparation or distillation of natural products? Nay, more the most peculiar and admirable results of art, if we thoroughly look into them, we shall find to be commonplace and inconsiderable, for indeed the very best of them have been discovered rather than invented. I especially refer to the printing press and the mariner's compass.
>
> That a magnet by its innate power should always point to the north is one of the greatest secrets of Nature; but it is hardly less wonderful that mankind, who has recognized the north by the help of this, has remained so long in ignorance about the other points of the compass, and how to turn this important discovery to account. . . . He who first discovered gunpowder was led to it more by chance than by study. And do not fiery and watery vapours, mingled somehow in the clouds, flash forth in lightnings more terrible and destructive than ever issued from a cannon's mouth? But we short-sighted, foolish

> mortals are ready to give almost divine honours to such discoveries, shallow though they be, even as the ancients made divinities of Bacchus and Ceres; but God Himself, the Creator, we quite shut out from His honours, and wellnigh from our thoughts and feelings.[18]

With such arguments, Joseph induces Politian to reconsider the claims of the new philosophy. Finally, in one lengthy discussion when the youthful skeptic raises many questions regarding the origin of the earth—whether it is due to fortuitous occurrence or to some natural properties lying hidden in its mingled elements—he is answered with Scripture and reminded that the idea that "all events are contingent or necessary is a new and rather puny philosophy." It is all very well, Gott says, to become adept in the chemistry of nature and to encourage useful and praiseworthy scientific experimentation, but not as those philosophers who "have been wont to let their studies end in the desire for knowledge and fame only." Such study should be solely for God's glory and it must always accept His unfathomable wisdom: "To penetrate beneath the surface of such mysteries as these is beyond man's power; they lie too deep; the labour is too great."

Gott's opinion is expressed in a two-volume, exhaustive view of life in the New Jerusalem, which is set in a romantic atmosphere with tales of adventure, love, and intrigue. He attempts, awkwardly at times, to integrate the description of the holy community with hair-raising episodes of bandits and narrow escapes from death. Philosophy is made to justify the Christian ethics that guide life in the ideal city, and the adventurous incidents are intended to teach resistance against temptation and immoderate living. Thus Samuel Gott's *Nova Solyma*, published many years after Thomas More's *Utopia* was conceived, still adamantly defends Christian virtue as the goal of progress; and though he accepts, even works for, advances in scientific explanations of the universe, he is fearful of the results. The ideal city basically must be founded on man's own virtue—improved and safeguarded—just as More, Eliot, and Hall had contended in the traditional approach.

The Natural Man

Simultaneously during the Renaissance, another group of writers was contending, perhaps more realistically, that man's character was not the basic problem but that institutions and customs of society were mainly responsible for the absence of utopia and therefore needed a general overhauling. Slight revisions would be inadequate. While this approach had its roots in Stoic and Epicurean philosophies, its reinforcement came at this time from the increasing interest in worldly, temporal matters and in the materialistic attitude that provided a basis for utilitarian values to come. From this point of view, the individual was seen in happier, more flattering lights. More's Utopians, it will be remembered, reflected this line of thought in two distinct ways: first, the Utopians' belief that man was essentially good and deserved happiness in the here and now of Amaurote, and, second, the attention Utopians trained on practical, economic, and social problems. Thus, although the Utopians' *major* emphasis was on the pursuit of virtue and this was their highest goal toward which progress was to be directed, they were neither self-flagellating in their pursuit nor did they overlook the necessities of reality in planning their economy.

Along the first line, namely, that human nature was primarily good and consequently one accepted it and even enjoyed the effort to know oneself, came François Rabelais in the high days of the French Renaissance and Michel Montaigne a generation later. Both inserted utopian sketches in their voluminous writings, and though differing sharply in details constituting the ideal society, their view of man as a worthy animal, noble at times, was similar. Rabelais, for example, was not in the least hesitant to trust the character of his aristocrats in the Abbey of Thélème. They were permitted to "Do as thou Wilt," and hence they led a charmingly gay and happy life, but one crammed with activities and intellectual interests. The Abbey was intentionally located 20,000 leagues from the typical Catholic monastery and equally far from the austere Protestant society in which one seldom heard laughter and never encountered opulence. Instead of the forbidding, cold, and constantly serious life of self-denial, Rabelais' noble ladies and gentlemen of high birth were all handsome and beautifully gowned; they even selected the color scheme for their attire according to that of the person whom they would accompany for the day. Their choice of jewels and dress was tremendous: raw silks, velvets, taffetas, damask, cloths of every imaginable color or woven in gold and silver, furs for trimming, pearls, and precious stones. All such "necessities" were furnished by craftsmen and traders situated in outlying buildings, for the Abbey was self-sufficient.

Remove the restrictions and man—at least one of aristocratic heritage which gave him a gentle and well-bred character—would happily fulfill himself on this earth. Place him in the midst of beauty with gardens, a fountain to the Graces, pictures and well-appointed apartments and he will flourish. Replace the monastic vows of chastity, poverty, and obedience with marriage, wealth, and liberty. No wall enclosed the Abbey or the hearts of its joyous people; they were free. Yet neither anarchy nor hedonism resulted. There was the rule of honor and conscience, of man's innate dignity and self-respect. And the hexagonal Abbey, six stories high with decorative towers, housed great libraries of texts in Greek, Latin, Hebrew, French, Italian, and Spanish (no English manuscripts were mentioned). The "inmates" were instructed so nobly that

> there was not a man or woman among them who could not read, write, sing, play musical instruments, speak five or six languages, and compose in them both verse and prose. Never were seen such worthy knights, so valiant, so nimble both on foot and horse; knights more vigorous, more agile, handier with all weapons than they were. Never were seen ladies so good-looking, so dainty, less tiresome, more skilled with the fingers and the needle, and in every free and honest womanly pursuit than they were.[19]

Rabelais (c. 1493–c. 1554), though a doctor by profession and an early student of anatomy and dissection, was not among the vociferous in pleading the benefits of scientific exploration; he, like More, was a little early for the issue. But he was most effective in reducing established beliefs *ad absurdum;* he lucidly displayed the ineffectiveness of the old scholastic curriculum; he directly took sharp aim at intolerance and bigotry with his facile pen; and spoofed the theological doctrines of guilt and damnation unless one be of the elect. There is a fresh breeze blowing through Rabelais' writings, clearing the air for free-questioning, trust in man's abilities, and belief in a friendly universe with no serpent lurking behind it. In this respect he joined More in his Christian humanist attitude; but when he portrayed the utopian life in Thélème he created a very different atmosphere—removing bonds of all sorts, giving his beautiful people a life of happiness and learning—and in his acceptance of human nature and desires, he was more realistic and more in accord with the actual attitudes society has subsequently expressed.

Montaigne, too, realistically accepted man for what he was, and spent his time and effort analyzing the nature, concerns, and interests of a human being—any or every man. He went beyond Rabelais' aristocratic selection of those relatively few individuals sufficiently noble in heart and self-disciplined in honor's code to enjoy Thélème's freedom. For Montaigne, nature in her plenitude gave to all men their share of vice and virtue, of foolishness and wisdom. He considered himself an ordinary man "except that he knew he was," and like Socrates, whom he greatly admired, he was a man who "desired neither to be nor seem anything else." From this philosophic point Montaigne started his chain of thoughts, examining dispassionately, doubting, and recasting his conclusions. His philosophy was genuinely progressive in its dynamic concept of change and evolution; and his goal could never crystallize. The process itself was of consuming interest and thoroughly empiric in its search, its consideration of negative and affirmative factors, its description of the subject or object as found under many different experimental conditions.[20] Method can scarcely be attributed to Montaigne, noted for his easy and frequent digressions, but his process of thought was frankly exposed in his essays and he purposefully aimed at truth, a relative truth for his subject matter: the nature of man and himself as an example.

One of Montaigne's particular targets, like Rabelais's, was intolerance, biased opinion, or a society's tendency to think itself superior to others. For this reason he condemned the Greeks' use of the word "barbarian" designating anything foreign to their own culture, and to prove his point he cited the story of a utopian people living in the newly discovered primitive country of Brazil. A man who had lived there for 10 or 12 years was Montaigne's houseguest, a simple, ignorant fellow and so more apt to give true evidence because he could not gloss his account in a sophisticated manner. Furthermore, the fellow had brought several sailors and traders whom he had known on his voyage to see Montaigne, thus corroborating the facts to Montaigne's satisfaction. The life of the Brazilian cannibals appeared to be of such purity, ruled by the laws of nature and only two ethical principles—valor against the enemy and love for their wives—that Montaigne regretted Lycurgus and Plato had not known of these primitive souls. They far surpassed "not only all the beautiful colours in which poets have depicted the golden age, and all their ingenuity in inventing a happy state of man, but also the conceptions and desires of Philosophy herself." Plato and Lycurgus would not have believed that society could live with such simplicity, "with so little human artifice and solder." And Montaigne imagines his explanation of their ways:

> This is a nation, I should say to Plato, which has no manner of traffic; no knowledge of letters; no science of numbers; no name of magistrate or statesman; no use for slaves; neither wealth nor poverty; no contracts; no successions; no partitions, no occupation but that of idleness; only a general respect of parents; no clothing; no agriculture; no metals; no use of wine or corn. The very words denoting falsehood, treachery, dissimulation, avarice, envy, detraction, pardon, unheard of. How far removed from this perfection would he find the ideal republic he imagined![21]

Montaigne continued his sketch of the primitive society, describing its temperate climate, its richly producing soil, the long buildings housing two or three hundred people, and their customs of eating one main meal in the morning and then drinking and dancing all day. This routine must have been interrupted when the cannibals waged war and tortured their victims according to the latest methods learned from the Portuguese who had landed on their shores. Clearly, Montaigne's purpose was twofold: one, he wished to expose the falseness and artificiality, the façade behind which European civilization was living; and two, he argued for the honest acceptance of man, his human nature, which combined the admirable and the ridiculous. In this he was a realist, and his essay *On Cannibals* must be read as ironic comment on the sophisticated life Montaigne knew only too well. He gave little thought to the advancement of scientific studies, banishing both letters and the science of numbers from his natural state; his concern was to advance man's knowledge and acceptance of himself. Thus Montaigne and Rabelais—at opposite poles in the cultural patterns they proposed—upheld the honorable tradition of man as a worthy creature, of man as the measure, and agreed with Thomas More at least in his humanistic opinion that man deserved happiness and might safely live by nature's reasonable rules. But this attitude wholly determined their utopian ideal, whether in the aristocratic or primitive state, and hence they claimed for their societies final goals radically different from the idealistic, Christianized virtue espoused by More's Utopians and their closer followers.

The Practical Man

Yet another element, more obviously realistic as we generally apply the term, was observed in the famous Utopian society: the attention devoted to practical, economic matters by the citizens of Amaurote. Many heirs holding this belief appeared and, of course, socioeconomic reform became a consuming interest in 18th- and 19th-century utopias when countless schemes reorganizing land control, the monetary system, and the economy were proposed. In the Renaissance transitional period the best example of this pragmatic approach to a better state was Robert Burton's utopia incorporated in the "Preface of Democritus Junior," as he styled himself when presenting *The Anatomy of Melancholy* (1621).[22]

Burton's projection of the ideal serves admirably to show the increasing secularization of thought, growing utilitarian standards, and realism in an objective analysis of society. These changes in intellectual and philosophic attitude prepared the stage for the entrance of the new scientific utopia, which Burton might well have written, but he did not choose to do so. Apparently he knew more "science" than many of his contemporaries who imagined its benefits and application. Earlier in the Preface, Burton anatomized both the human body and the medical knowledge of the age; later in the *Anatomy*, especially in the section entitled "Digression of Air,"[23] he revealed his reading of the well-known works that laid the groundwork for significant discoveries. For example, he recognized Copernicus' hypothesis, upheld in "sober sadness" by Digges, Gilbert, and Kepler, and said, "If it be so that the earth is a moon, then are we also giddy, vertiginous and lunatic within this sublunary maze." Burton was familiar with Galileo's works, including the *Sidereus Nuncius* or *The Starry Messenger*,[24] on which he commented that Galileo had seen the Milky Way comprised of little stars "like so many nails in a door." Moreover, he fully realized the contribution of Greeks like Aristarchus, Pythagoras, and Democritus to scientific thought and mentioned the revival of ideas in Copernicus' hypothesis. Burton was extremely well informed for a layman of the early 17th century and welcomed the new theories being advanced.

He specifically mentioned Campanella and Bruno together with their predecessors who believed in an infinity of worlds; he had read Campanella's *De Sensu Rerum*, pointing out its agreement with Kepler's notion that the planets were inhabited; and he knew of Campanella's apology for Galileo. When he spoke of the "irregularity" discovered in what had been considered the perfect circular heavens, he was not perturbed, but he also did not take many final positions regarding the new facts. Although he confidently counted "upon the eternal movement of innumerable worlds," the conclusion, as always with Burton, was based on authorities cited and the suppositional "if they be correct." In attitude he remained the long-winged hawk as he pictured himself, roving over and around the earth, viewing from a great height the various theories of the operations of this world and others with eager but dispassionate curiosity.

In his one play, the satirical *Philosophaster*, Burton allows his bona fide professors to question a fraudulent scholar, Theanus, in this area of knowledge:

> *Polumathes.* What thinkest thou of the threefold movement of the earth? What dost thou think of the new star—is it sublunary, or of the Heavens?
>
> *Philobiblos.* Raimerus hath fetch'd forth a new hypothesis, Tycho Brahe a new one, Fracastorius a new one, Helisaeus Roeslin a new one, Patricius a new one, Thaddeus Haggesius, the Hegeck Doctor, a new one.
>
> *Theanus.* Lord save us! what kind of incantation is this?
>
> *Polumathes.* All these reject the element of fire.
>
> *Theanus.* Harping on that again!
>
> *Polumathes.* Maginus inveigheth against all these, and hath lately made out eleven Heavens; herein, belike, he offendeth against the first principles of mathematicks: opticks on the one hand, philosophy on t' other. Whom, then shall we believe? To what conclusion hast thou come? I entreat thee earnestly.
>
> *Theanus.* What's above is naught to us.[25]

It is the foolish Theanus who cares "nothing for astronomy," while the sound doctors have raised questions presented by various 16th-century authors on astronomy. Although the play was written in 1606, revised in 1615, and acted two years later at Christ Church, Oxford, there was no change in this attitude, and Burton continued to hold the same questioning position later in all six editions of the *Anatomy* from 1621 to 1652. Still he knew the issues, admitted that fire, the highest of the old four elements, had been quenched by the new learning, and obviously he was conscious of the upset caused by the appearance of new stars in the formerly "fixed" heavens.[26] Burton was also reputed to have been an "exact mathematician,"[27] and the opening lines of his play reflected this interest. The impostors were rehearsing geometric and algebraic terms calculated to display their erudition and so deceive the gullible public.

Furthermore, in his eclectic and omnivorous reading Burton had encountered the current "scientific" utopias of his peers. By the time he revised his *Anatomy of Melancholy* for the 1628 edition, he had become acquainted with Johann Andreae's *Christianopolis* and Bacon's utopia, and by 1638 he included Campanella's work. Knowing that these utopias of the new order did not affect his attitude, he specifically dismissed them as "witty fictions."

> Utopian parity is a kind of government to be wished for rather than effected, *Respub. Christianopolitana*, Campanella's City of the Sun, and that *New Atlantis*, witty fictions, but mere chimeras.[28]

And for slightly different reasons he goes on to reject Plato's community which, he says, may be severely criticized for many things, but primarily because it "takes away all splendour and magnificence." In disagreement with Plato, Burton keeps a monarchical government and allows degrees of nobility based on heredity.

Thus it is neither in Plato's *Republic* nor in the contemporary utopias of Burton's time that parallel interests or possible influence may be found. Instead, it is Sir Thomas More's *Utopia* that bears the closest resemblance to Burton's "poeticall commonwealth," and the similarity lies mainly in their joint concern for practical solutions, social, and economic reform. The ills of mankind are fully set forth for the reader in the "Preface of Democritus Junior," and the imaginary utopia that follows is planned to solve these ills—very much as More had related the second book of *Utopia* (containing the ideal state) to the first, which described the problems of society, England in particular. And, like his predecessor, Burton had little hope that his ideal could be realized. In truth, he says, "These are vain, absurd, ridiculous wishes not to be hoped." Even the words are reminiscent of More's acceptance of the fact that one "may rather wish for than hope for" his dream world to come into being. In this same realistic spirit Burton gibes at the possibility of actual reform: what the world really needs, he suggests, is "a just army of Rosy-cross men, for they will amend all matters (they say), religion, policy, manners, with arts, sciences, etc.; another Attila, Tamerlane," or a Hercules who might transport himself by the power of a magic ring to any part of the world and reform all distressed states and persons. Since such aid is unlikely to come and the task is really beyond Hercules' talents, the author proposes to employ the poet's license, so he tells us, and "build cities, make laws, statutes" for his own pleasure.

Burton simply enjoyed himself in dreaming of a perfected state. He did not care where it was located and never, in all the editions, made up his mind. It might be in *Terra Australis Incognita* (the unknown continent was a favorite spot for utopias) or on a floating island in *Mare del Zur*, or one of the Fortunate Isles, "for who knows yet where, or which they are?" Or there is plenty of space in America and on the northern coast of Asia. So he considered several possibilities and left its placement in mid-air. He bothered with no informant, no trusty friend to tell him what he saw. Consequently, there is none of the fictional framework expected in the literary genre; Burton's work is frankly a figment of his own imagination.

Yet his dream was amazingly full of common sense and surprisingly adaptable to practical circumstances. For example, he insisted on standardized measures and rectified weights, the advantages of which he might well have seen personally in his three-year term as Clerk of the Market. This was a real problem in England at the time and had already aroused the interest of Francis Bacon. In a speech delivered in 1601, Bacon had said,

> I'll tell you Mr. Speaker . . . that this fault of using false weights and measures is grown so intolerable and common, that if you would build churches, you shall not need for battlements and bells other things than false weights of lead and brass.[29]

Similarly, Burton's acceptance of usury was notable for utopias, unusually expedient, and generally indicative of his thinking:

> Brokers, takers of pawns, biting usurers, I will not admit; yet because we converse here with men, not with gods, and for the

> hardness of men's hearts, I will tolerate some kind of usury. If we were honest, I confess, we should have no use of it, but being as it is, we must necessarily admit it.[30]

In this way he adjusted to reality—men cannot be gods—but he firmly put usurers and borrowers under control of supervisors and limited interest rates. When considering such business transactions and regulations of trade, he reflected always his concern for the less fortunate. He would have taxes only on luxury items, correct the problem of enclosure in his improved England, abolish private monopolies, care for the genuinely needy at state expense, and banish idle, lazy rogues.

In the land he imagined, no barren acre existed, "not so much as the tops of mountains: where nature fails, it shall be supplied by art." Each acre was planted according to soil analysis and the process of tillage was carefully supervised. Man's art could apparently perform miracles, and anyone who "invents anything for public good in any art or science, writes a treatise, or performs any noble exploit at home or abroad, shall be accordingly enriched, honoured, and preferred." In spite of such evidence of faith in man's ability and the realistic emphasis on practical problems, Burton was not thinking like the new utopists. His marginal note for the passage just quoted shows that his thought was patterned on Matthew Ricci's account of China in the 16th century.[31] The gloss on Ricci reads: "In guiding public affairs, only the liberally educated are admitted and they do not require the favor of an official or of a king; everything depends upon the demonstrated knowledge and competence of the individual." When Burton recommended that honors and preference be given to those who invented something of value or wrote a treatise of worth, his prodigious memory could easily have recalled Aristotle's remarkably similar words (*Politics* 2.8) when he reported that Hippodamus' plans for a better city provided that "those who discovered anything for the good of the state should be honoured." Furthermore, in the same work (7.17) and elsewhere, Aristotle had also stated: "The deficiencies of nature are what art and education seek to fill up." So, too, Burton's concept that art completes nature's work when necessary reflected the old purposive universe; he was still speaking the language understandable in Sir Thomas More's town square of Amaurote.

Although Burton suggested basic changes in educational techniques and curricula, he was neither an innovator nor sufficiently detailed in his recommendations to be judged unique or especially forward-looking. It is true that grammar and languages are "not to be taught by those tedious precepts ordinarily employed, but by use, example, conversation, as travellers abroad, or nurses teach their children." Colleges will exist for mathematicians, alchemists ("not to make gold, but for matters of physic," Burton explains in a note), physicians, artists, and philosophers, so that the arts and sciences may be sooner perfected and better learned. But there is no hint in Burton that the glorious future for men will come primarily from scientific developments, and his utopia does not depend to any great extent on the results of laboratory experiments and research.

He did not share the enthusiasm and hope characteristic of Andreae and Bacon, or predicate society's improvement upon the new knowledge and the conquest of nature. In his own life, Burton was neither a reformer nor a decisive man of action like his fellow creators of utopian plans. He preferred the sedentary pursuit of learning and seldom left the walls of Oxford except for an occasional stroll along the waterfront to enjoy the scene and language of barge-hands, just as he believed his predecessor Democritus had found pleasure in such walks. In the Preface, Burton remembers from his reading that the ancient philosopher would sometimes bestir himself from his studies and private life to " 'walk down to the haven, and laugh heartily at such variety of ridiculous objects, which there he saw.' Such a one was Democritus." Such a one also was Burton who, in many particulars, seems to have patterned his life after the Greek philosopher's whom he so greatly admired. From his model, Burton could easily derive endorsement for his materialistic, secular approach to the problems of society, because Democritus as a forerunner of the Epicurean school represented the atomic theory of matter in which atoms (solid and indestructible) together with space or void between them accounted for all the forms and qualities that we recognize in things. Birth and death were only a rearrangement of atoms; matter was not lost but took new form. Democritus was further reputed to have been a patient observer and to have practiced dissection; in comparable fashion Burton anatomized melancholy, frequently including his careful observations on human nature and behavior. So Burton, or Democritus Junior, appears to have been a quiet, rather withdrawn man, yet remarkably cognizant of worldly affairs, and most attentive to the problems of humankind.

In deciding which features his utopia would adopt, he frequently compared his plan to practices existing elsewhere; the whole scheme was grounded in actuality. Just as he accepted the fact that we are not gods, but men with the frailties common to human nature, so he recommended on a practical basis a renovated society, not so perfect as he might have imagined, but one that feasibly could be achieved. The outstanding element was his concern that every man have a decent living standard, that the poor and oppressed receive their fair share, though all must labor diligently for their rewards. No utopist had gone so far in projecting a socioeconomic program as the means to a better state. Professor Patrick has summarized it well: "Burton's outlook was materialistic and utilitarian. . . . He was a pioneer in advocating a secular, non-communist, planned society."[32]

Dr. Patrick further credited Burton as "the first utopist to use scientific comparative procedure as the basis of his sociological theories: in this respect he anticipated modern sociological method." Certainly Burton often justified his inclusion of one or another element in his society on the grounds that such a practice existed elsewhere as, for example, when he cited the Chinese examinational system as the basis for selecting officials in his ideal state. In other instances he compared his recommended program to that of England or Holland, and he anatomized problems of actual states to find his better procedure. This realistic approach was unusual for idealistic utopists, though it had been employed by Aristotle in his *Politics* (Book 2) when he examined with derogatory criticism the imaginary plans of Plato, Phaleas, and

Hippodamus as well as the best existent states: Sparta, Crete, and Carthage. Of course, the *Politics* does not qualify as a utopia, so it cannot challenge Burton's award of a first; for the same reason Machiavelli's use of comparative examples does not displace Burton from first position. But the process itself is not new; it is merely its use in utopian writing that is unique. The important question here, however, is the relationship of the comparative approach to the appearance of "science" in utopia. Burton was orderly and objective in analyzing his materials; he dissected both society and the individual afflicted with melancholy in his *Anatomy*, but he still invoked the authority of the written word to substantiate his findings. His respect for the printed dictum far surpassed his trust in observation. While his basic search of phenomena was extensive and often comparative, his main purpose remained descriptive; in fact, he seldom reached final conclusions of any sort. In his utopia, of course, he was forced to decide, to define the ideal living conditions, and often he adopted customs employed in other societies but, when he mentioned them, it was in terms to justify his choice, claiming their effectiveness elsewhere as proof.

The man who took this analytic process further and employed it more obviously in framing a utopian state was Burton's younger contemporary, James Harrington (1611–1677). In the *Oceana*,[33] Harrington examined the Commonwealth of Israel, the Roman Republic, and the modern political structure of Venice, sifting the problems of each to find causes and effects, the origins of disorder, before deciding upon the component parts of his projected state. His was a political anatomy that led to a basic conclusion without support from quotable, corroborating authority. A close and careful investigation of the facts of social history led Harrington to his principle: in the balance of economic property (which then meant agrarian power) lay the balance of political power, the stability and strength of government. In making his "discovery" Harrington had used the experimental method of anatomy, had built upon a comparison of facts, rather than deductive reasoning. Instead of starting philosophically from the nature of man and deducing the best form of government for the human creature—a process for which he criticized Hobbes severely—Harrington attempted to determine empirically the universal laws operative in society. In this effort he reflected the atmosphere of his age and the general search for new knowledge in all areas.

R. H. Tawney, in a lecture delivered before the British Academy, heralded Harrington's approach in these terms:

> His central conception, that institutions are not accidental, or arbitrary, or susceptible of change at will, but are the necessary consequence of causes to be discovered by patient analysis, was all in the spirit of the New Learning of the day.[34]

And men in his own day saw Harrington's work in the stimulating light of new discoveries. In the first collected edition of his works published in 1700, John Toland, his enthusiastic editor, rapturously exclaimed:

> That Empire follows the Balance of Property . . . is a noble Discovery, whereof the Honor solely belongs to him, as much as those of the Circulation of the Blood, of Printing, of Guns, of the Compass, or of the Optic Glasses to their several Authors. 'Tis incredible to think what gross and numberless Errors were committed by all the Writers before him.[35]

Harrington's *Oceana* could have suggested the extravagant metaphor to Toland, for there the parliament was "the Heart, which, consisting of two Ventricles, the one greater and replenish'd with a grosser matter, the other less and full of a purer, sucks in, and spouts forth the vital Blood of *Oceana* by a perpetual Circulation."[36] And Harrington justified movement in government by pointing out the motion of the universe (whether the earth or heavens move) which yet endures with stability. Regardless of his imagery and understanding, Harrington contributed no scientific facts. He belongs, as he has been placed, among the fathers of political science along with his admired Machiavelli and others, but not in the group of new utopists placing science as a study in the classroom and building laboratories for scientific discovery. Oceana's schools offered young people a curriculum in "mechanicks" by which Harrington meant husbandry, manufacturing, and merchandizing. His universities remained devoted to the training of divines and statesmen. Science had not revolutionized society in Oceana any more than in Burton's earlier, ideal society, but it had definitely affected both authors' approach, their method of planning and evaluating the constituent elements for utopia.

The Renaissance was a time of many models, new ideas, and change. Inherited classical works offered more than patterns; they suggested and stimulated new concepts and ways of thought as well. Along with the idealistic tradition, which expressed itself in religious and secular utopian writing, had come the realistic and expedient strain of thought from the Greeks and Romans, which had not constructed utopias in the earlier period but which now had entered and caused sharp adjustments in the utopian environment. This naturalistic outlook spread and invaded many areas of thought, creating an atmosphere conducive to questioning and experimenting, to the growth of scientific learning which, in its turn, affected the method of utopists like Harrington, who came later in the period.

As the philosophic approaches to utopia widened during this transitional period, many doors opened leading to new goals for the ideal society. With the general temper of the age came more practical, utilitarian standards and objective, clear-sighted appraisals like Burton's. Sir Thomas More had suggested the importance of practical problems, but Burton had gone well beyond the notions of the 16th-century chancellor of England, and he certainly did not emphasize that other main facet in More's *Utopia:* Christian virtue. This latter goal for utopia was gradually losing ground as man's thoughts turned more and more to daily needs, an easier life for all, and ways to obtain materialistic comforts. Still the goal of virtue was not without champions: writers such as Joseph Hall, Sir John Eliot, and Samuel Gott upheld the long, idealistic tradition established in Plato's *Republic.* Even after scientific pronouncements had startled and excited men, they maintained that utopia lay in the pursuit of right values alone.

On the other hand, Rabelais and Montaigne contended that

man need not remodel himself completely; his own human nature was a trustworthy guide, and enjoyment of life's pleasures was his natural right. Within this more realistic framework Robert Burton expressed his timid hopes for social and economic reform and James Harrington, one of many political utopists[37] of the period, saw the main source of man's betterment in a form of government actually patterned on economic forces.

In the first quarter of the 17th century, however, another group of utopists had come to the conclusion that the new age of mankind was to be the result of science and the new learning. While their visions did not instantly blot out all other views of utopia—we have seen the various solutions of their contemporaries—still their quick and creative adoption of scientific discoveries proved to be the main road to the future that we have traveled. Here are Andreae, Bacon, Campanella, and others who set the stage on which George Orwell and Aldous Huxley have drawn the curtains, leaving man controlled in the darkness of his own creation. In the period of scientific gestation, there was no thought of such a conclusion; the new utopia was flooded with the brilliant light of new knowledge.

IV

THE NEW UTOPISTS

AT THE BEGINNING of the 17th century the heavens and the earth were undergoing a tremendous transformation not only in the mind, but in the very eyes of man. A vision of distant, possible worlds was actually seen by Galileo through his optic tube, the telescope, and his *Starry Messenger* announced the reality of many stars hitherto unknown and hence unnamed. After Galileo saw the "small stars set thick together" in the Milky Way, it became more than the old glorious highway of the gods. Not only were Hermes' wings clipped by the new messenger, but the angels too required reassignment: they were no longer personally needed to "drive the planets around the sky."[1] Johann Kepler, instead of being lulled by the beautiful music, the harmony from Plato's and Pythagoras' spheres, had examined and accurately explained their movement. The perfect circle had been forced to give way before the ellipse. Other outstanding students like the Dane, Tycho Brahe, and Giordano Bruno had made their contributions to astronomy and the new view of the universe in the latter part of the 16th century, laying the groundwork so that now definitive statements with far-reaching results could be made. The evidence was accumulating in favor of Copernicus' theory that this world was spinning in space, though the issue was not conclusively settled until Newton's work toward the end of the century.

While astronomical discoveries were the most striking in impact and came with astonishing rapidity in the early decades of the century, basic facts were also being revealed in other closely related and important areas. William Gilbert, personal physician to Queen Elizabeth and an early advocate of one part of Copernicus' views, presented in 1600 the first major, original contribution to science published in England.[2] Translated from Latin, the title read: *On the Magnet and on Magnetic Bodies and concerning that great magnet, The Earth, a New Physiology*. And, in what we think of today as more properly "physiology," William Harvey described for the first time correctly the circulation of the blood. He had formulated his concept by 1615, but did not publish the demonstration until 1628. Significantly in this work, Harvey applied principles of measurement (as when he determined the capacity of the ventricles of the heart) which he may have learned when studying at Padua (1600–1602) while Galileo was teaching there.[3] Developments in mathematics and mechanics were great and implemented progress in many fields, both practical and theoretical. They made possible the growth in quantitative studies that could finally and factually defeat the qualitative world based on abstract values, which was the 17th century's general inheritance.

Such discoveries, of course, do not suddenly appear; the scientists working at the beginning of the century were well aware of their heritage and often, in their writings, stopped to argue or agree with their scientific forefathers. During the Renaissance another strong chain of ideas about the physical universe had been woven, linking the clever Greeks and some outstanding figures of the medieval period to the modern world. Copernicus indicated his awareness of the source of his idea when he stated in the introduction to *De Revolutionibus* that he had found initially in Cicero that Nicetas thought the earth moved, and later he had discovered through Plutarch that there were others who held the same opinion. No doubt to bulwark this revolutionary doctrine, being published for the first time in full, Plutarch's words were copied out as follows:

> Some think that the Earth is at rest; but Philolaus the Pythagorean says that it moves around the fire with an obliquely circular motion, like the sun and moon. Herakleides of Pontus and Ekphantus the Pythagorean do not give the Earth any movement of locomotion, but rather a limited movement of rising and setting around its centre, like a wheel.[4]

The conclusion drawn is that having so encountered the notion, he decided to test whether or not, on this theory, "demonstrations less shaky than those of my predecessors could be found for the revolutions of the celestial spheres."

Similarly, Kepler's thought was stimulated by the Pythagoreans' idea of five regular solids, which were later called the "Platonic bodies" and with which Kepler started his unitary system explaining the structure of the universe.[5] Bruno had accepted not only Copernicus' hypothesis but he also knew the writing and thought of Cardinal Nicholas of Cusa (1401–1464), who believed the earth might move and the world be infinite, though he formed no systematic astronomical theory.[6] And Galileo notably started his research from a detailed study of the mechanical doctrines of Aristotle, several of which he proceeded systematically to disprove. Thus one man builds upon another's work in the long sequence of amassing knowledge, but in this instance of world history the rediscovery of the ancients during the Middle Ages and the Renaissance had greatly accelerated the process. The newer men tended to return directly to the original thinker, either arguing against or building upon his work; they less frequently dealt with the go-betweens and interpreters.

Yet this chain of scientific thought had neither significantly influenced utopian trends nor the populace at large. Galileo's *Starry Messenger* was one of the important literary works that mended this dichotomy. He wrote for the layman and told him that the shadowy region of the moon was marked with spots "like a peacock's tail with its azure eyes," that it resembled a vase hot from the kiln plunged into cold water, which had given it the crackled surface of frosted glass. In this choice of metaphor, Galileo revealed plainly that he came from the region of Venice, famous for its glass-making, and he knew his immediate reading audience would recognize what he had seen through his tube. His desire was to give many others his knowledge. And so the *Starry Messenger* brought more than news of planets and stars to this world: it announced to men as well what was going on in the scientist's world. In comparable style William Harvey presented his first explanation on the motion of the heart and blood, so that any interested reader could easily understand. He said that the heart was as the sun of the microcosm, "even as the sun in his turn might well be designated the heart of the world," that it was like the "household divinity" which nourishes and quickens the whole body. He called upon the butchers to bear witness to the speed with which the body of an ox is drained of its blood. Such writing was for all literate persons who would naturally be interested; specialists writing technically for their narrow audience had not yet emerged in significant numbers.[7]

Fortunately in the earlier days of the century many curious and active minds received the message from Galileo and the writings of Kepler, Gilbert, and Harvey. The reception afforded the new knowledge, however, varied markedly. As has been observed, Robert Burton was thoroughly familiar with the *Starry Messenger* and current discoveries as well as the long chain of development from Aristarchus, Pythagoras, and his favorite Democritus through the now almost forgotten names of Fracastoro, Roeslin, and others of the 16th century. Yet Burton remained calm and personally uninvolved, quoting extensively from his prodigious knowledge, but not injecting the new ideas into his utopia. Joseph Hall and Samuel Gott, too, knew at least generally the novel currents of thought and cautioned against their adoption: the ideal society had better look to its character and ethical standards instead. In sharp contrast, Andreae, Bacon, and Campanella—living in different countries but in the same period of time—saw the challenge in the new thought and quickly translated it into utopian terms. If science was entering its modern phase, so was utopia.

These progenitors of the modern utopia were not scientists actively engaged in research, although a few sporadic attempts at something of the sort were made by Francis Bacon. Unlike their contemporaries—Galileo, Kepler, and Harvey—they left the world no important discoveries or new natural laws. They were amateurs in a century when amateurs were a most potent factor in spreading the new knowledge and gaining its acceptance.[8] They started from such works as the *Starry Messenger* and extended the doctrine, interpreting its meaning and effects on human society. In those days before the body of scientific knowledge was organized into distinct, highly specific fields of study, the intelligent man of general interests could contribute a great deal. Not the least of the new utopists' contributions was simply the enthusiasm with which the new facts were greeted and the amazing speed with which science, still in its elementary period, was catapulted into a major role in the life of man. Such lively projections of values expected from scientific work not only bred more utopian statements, but furthered the growth of interest among the reading public and the more influential citizenry.

Excitement over the new learning, as it was often called, was the main characteristic possessed by the new utopists in common. Aside from this, they were rather odd bedfellows: Johann Valentin Andreae (1586–1650) was a German Lutheran minister and teacher, Tommaso Campanella (1568–1639) was an Italian Dominican friar, and of course Bacon (1561–1626) was the Lord Chancellor of England and a noted man of letters. Regardless of their various professional labels, they were men of unusually wide interests even in their own time. With Bacon, they took all knowledge to be their province and worked industriously at the huge task. Bacon included this goal as the third division of his Great Instauration in which he announced the plan to examine "the Phenomena of the Universe; that is to say, experience of every kind and such a natural history as may form the foundation of philosophy." Narrow categories of interest, the professions into which we fit people today, would have been unthinkable to these avid minds of the late Renaissance. Just as Galileo was a famous teacher at the University in Padua, and Harvey, Gilbert, and Copernicus were all practicing physicians, so the utopian authors were men of several parts. While the focus must be on their utopias, it is necessary to see them at least from a broadside view in order to evaluate their thought. The primary questions are what they actually knew of science, and how it generally entered their utopias, thereby affecting the literary genre as well as their disciples who continued the pattern. Subsequent chapters deal with details of issues, method, and the results pictured in the ideal scientific society.

Although no one of the three progenitors qualifies as an outstanding contributor to scientific discovery, each was well informed and generally knew the experimental work and findings of the scientific leaders of his day. In several instances they were in direct correspondence with the men making the great pronouncements on natural law. And there is indisputable evidence that the new utopists held their forward-looking scientific attitudes often before they knew the work of their contemporaries in the laboratory—which means, at least, that their attitudes and opinions developed from earlier Renaissance contributors or, at best, the utopian author himself decided upon the course to be taken to reach scientific truths. In either case, he was ready to receive the new ideas; and he was intellectually able to understand and evaluate the new findings.

Tommaso Campanella

Campanella, a Calabrian from Stilo, was the first of the three to draft a utopia: his *Città del Sole* appeared initially in manuscript form in 1602.[9] In his earlier years as a student, Campanella had studied the medieval Dominican doctrines of Thomas Aquinas and Albertus Magnus, which led him to join the order in 1582, when he was just 14 years old. Having been mentally nurtured on the Aristotelian tenets of the older theologians, he must have experienced a profound shock when, by chance soon after entering the order, he encountered the writings of Bernardino Telesio, the renowned anti-Aristotelian and experimental philosopher. Telesio's works had a great and lasting influence on the young student's thought, even though his superiors in the cloister at Cosenza immediately frowned upon his interest in the new ideas and, in fact, they prevented Campanella from ever meeting the great Telesio, who had established his academy and was residing in the same small town. It is pathetic to recall the story told of the young man's homage on the death of Telesio in 1588, when the body was displayed upon the bier in church and Campanella prayed beside it, placing poems upon the body.[10] One is free to imagine (since the dates are uncertain)[11] that the poems included his own, entitled *To Telesius of Cosenza*, which begins:

> Telesius, the arrow from thy bow
> Midmost his band of sophists slays that high
> Tyrant of souls that think; he cannot fly:
> While Truth soars free, loosed by the selfsame blow.[12]

It matters not if the sonnet belongs with those Campanella composed during his later years of imprisonment for his beliefs. The point remains that he held staunchly to the teachings of Telesio and bravely continued throughout his life to shoot the arrows of free thought directly at its greatest enemies as he saw them: the Sophists, "Traitors to thought and reason, jugglers blind;" the Hypocrite, "entwined with lies and snares;" and the Tyrant, "wearing the glorious show of nobleness and worth," who "keeps you confined."[13] His poetry is powerful, terse, often rough with an unchiseled quality that adds to its force. And apparently his personality was similarly forceful and dominant, possessing as a result the magnetism of leadership that caused the church and state alike to fear its effects.

His first major philosophic work written in the year following Telesio's death, the *Philosophia Sensibus Demonstrata* or *Philosophy Demonstrated by [the evidence of] the Senses*, was a defense of the master's teaching and led to Campanella's first appearance before the Dominican tribunal and consequent imprisonment for one year. Despite his early introduction to the confinement that eventually lasted more than 27 years, including 8 in an underground dungeon and unbelievably horrible torture on three occasions, he persisted in promoting the new theories. By the time he was 25 years old, he had written two others of his best known works: *De Sensu Rerum* and *De Investigatione Rerum*, his contention being that man can understand the world only through the senses, that all philosophic knowledge then must be based on sensation and the direct study of nature. Thus Campanella with Telesio looks back to Lucretius' belief that all knowledge was derived from the senses, as well as forward to the increasing importance of the theory. On this line alone one can understand why Bacon heralded Telesio as "a lover of truth" and the "novorum hominum primus," first of the new men, the inaugurators of modern thought.

Telesio had rejected the authority of abstract reasoning together with various Aristotelian theories and, in one instance at least, he made an important advance: he replaced Aristotle's "form" by the concept of "force" although he was not very definite about what this meant.[14] The force was comprised of two opposing and attracting elements, one of expansion or heat (sun) and the other of contraction or cold (earth), and their interaction accounted for all the diverse forms and types of existence. Motion was thus an operation of heat, not new heat produced, but pre-existing heat excited.[15] In Telesio's theory, specifically, is found a source for the strange notion held by citizens in Campanella's *City of the Sun* when "they assert two principles of the physics below, namely that the Sun is the father and the Earth the mother." It must be remembered that in the preparatory period before the new mechanical universe was presented, many and varied conceptions of the world and its phenomena were of course proposed. Men were seeking for concepts that would accommodate the facts appearing, yet fit with some old beliefs that they were loath to leave. They sought an over-all structure to replace the beautifully consistent medieval scheme. The significance then of Telesio's philosophy, primarily his general emphasis on the Interrogation of Nature and his dependence on data given by the senses, is that of a forerunner of modern empiricism. He himself did little observing and recording; he continued theorizing about nature, but it set off free thinking, and southern Italy produced two followers in Campanella and his brother in the Dominican order, Giordano Bruno, who died for philosophical liberty.

In the northern area, the giant figure of Galileo was dominating the scene and, only a year after his report by the *Starry Messenger* (1610) on the heavens, Campanella wrote him in considerable excitement, saying that he believed the moon was inhabited and wondered whether the inhabitants were "of the blest or mortal like ourselves." He proclaimed Galileo the glory of Italy who was to surpass all astronomers even as Vergil and Dante had displaced Homer: "Amerigo gave his name to a new earthly world, thou wilt give thine to a new celestial one." And Galileo was urged further to formulate a universal philosophy, not to confine himself only to physical science.[16] Fortunately, Galileo did not launch into metaphysics but stayed with his "limited" field, which proved rich indeed. When the Holy Office first censured his work a few years later in 1616 and placed the injunction of silence upon him, Campanella immediately wrote an excellent defense, the *Apologia pro Galileo* (published in the freer presses of Frankfurt, Germany in 1622). This tract, written in prison by a man who had been forced at one point to feign madness constantly for over a year in order to save his life, is one of the best and most courageous documents existing in defense of the Copernican system and man's right to freedom in such investigations of natural phenomena. It will be considered in detail in the chapter on answers to religious authority; the point here is to recognize Campanella's general position and his appreciation of the great Florentine's work. He remained in correspondence with Galileo and in 1632, just before the final condemnation, he offered to defend him before the Holy Office. Such an audacious act could only renew suspicion about Campanella, who had been finally released from his life sentence of imprisonment only three years earlier. When, in addition, a conspiracy in Naples was started by one of Campanella's disciples, he was forced—with the help of the Pope this time—to escape to France, where he spent his last few years working to compile a complete edition of his writings and still agitating for a renovated world.

He wrote about a hundred works which included, besides his philosophic, political, and religious writings, treatises on medicine, mathematics, physiology, fortification, rhetoric and poetry, interpretations of dreams, and some six volumes on astrology. It was the truly prodigious production of a Renaissance man. His writings vary in their value, of course, but the peaks of important contributions are there. He was not unappreciated in his own time: one contemporary considered him a philosopher of "extraordinary gifts, skilled in mathematics, astrology, medicine, and other sciences."[17]

Among Campanella's most widely read works is the utopian

City of the Sun in which his various beliefs are incorporated into one highly unified view of society. Following his contention that knowledge comes to man via the senses, he has built the entire city on the plan of seven concentric walls on which all—he claims all—known phenomena are vividly portrayed; the very structure of the city literally embodies complete knowledge. His citizens, the Solarians, on a daily errand could easily absorb both the history and latest developments in any area of research, and children in a systematic educational program studied the walls. For example, when minerals, metals, and stones are pictorially shown on the inner side of the second wall (both sides of walls are functionally used), a little piece of the matter itself is there with suitable explanation. Or when liquids—lakes, seas, rivers, wines, and oils—are displayed on the outer surface of the same wall, a small carafe containing the actual liquid is built into the wall. Sources of extraction, strength, and qualities are explained. It is thoroughly carried out, and the visual exhibits progress along the old chain of being from the stones and lowest matter to plants, fish, birds, animals "beautifully and accurately delineated," and finally man together with his inventions. The first wall is devoted to mathematics and the sixth to "all the mechanical arts" with their inventors.

While Solarians have not reached astronomical conclusions, at least they are open-minded:

> They praise Ptolemy, and they admire Copernicus (though they put Aristarchus and Philolaus before him). Yet they say that the one does his counting with pebbles and the other with broad beans and neither of them uses the very things being counted. Consequently they pay the world off with play money instead of gold. However, these people are very interested in matters of this kind and study them very closely, for it is important to know how the world is constructed, whether it will end and, if so, when, what the stars are made of, and who inhabits them.[18]

Campanella was fair in his evaluation of the merits of these systems, the Ptolemaic and the Copernican at this time, just as he recognized Copernicus' predecessors in Aristarchus and Philolaus, who both believed the earth like other planets was revolving around the sun. As a matter of fact, the Copernican hypothesis retained much of the ancient theory and was basically a conservative rearrangement to reduce the number of Ptolemaic circles to 34. So Copernicus did use "broad beans" compared to Ptolemy's "pebbles." It was further true that neither of them effectively used "the very things" being counted. Copernicus was no systematic observer; his conclusions were drawn for the most part in his study.[19] Ptolemy had used observation, such as it was in his day, and he had devised some instruments to aid his calculation, but both men, of course, lacked the instrumental means necessary for close and continuous observation. Furthermore, mathematically it made no difference whether the earth or the sun moved. Thus, until Kepler made fundamental changes in Ptolemy's details to accommodate the more accurate observations recorded by Tycho Brahe,[20] the scales were balanced between the two views. Solarians, speaking for Campanella in 1602, were merely cautious in making the final decision, awaiting the proof, but dispassionately and intelligently considering the issue nevertheless. They were too early also to have had Galileo's report on the heavens to which Campanella so enthusiastically replied in his letter of January 13, 1611.[21] It is interesting to notice that later when Campanella was defending Galileo's pronouncements, he especially commended him in six passages[22] for his description of natural phenomena "according to the testimony of observation;" he had demonstrated "the truth of his doctrine by sensory observation." This was what Campanella's Solarians correctly found missing in the work of Ptolemy and Copernicus.

Galileo's discoveries did more, however, than endorse the theory of the sensory test believed in by Campanella: they, together with Tycho's observations, upset some of the basic tenets of Campanella's philosophy as he had stated it when a young man. He admitted as much in the introduction to the final chapter of the *Apologia pro Galileo:*

> For many years I regarded heaven as fire, itself the fountain of all fire, and the stars to be constituted of fire, which Augus-

tine, Basil, and other Fathers had maintained, and recently our Telesio. I also attempted in the *Physical Questions* and *Metaphysics* to refute all the arguments of Copernicus and the Pythagoreans. Yet, after the observations of Tycho and Galileo proved conclusively that new stars are in the heavens, comets are above the Moon, and that spots move about the Sun, I suspected all stars are not composed of fire. . . . My doubt again was increased by the argument, in contrast to what we had been taught, that the sphere of fixed stars could not traverse so many thousand miles in a single moment. The Medicean and Saturnian stars which move about Jupiter and Saturn permit neither a single Sun; nor the center of love, that is, the Sun; and another of hate, that is, the earth; as we say in natural philosophy. Finally, the similarity of many external qualities of the fixed stars to the external qualities of the planets caused me to consider carefully the opinion which Galileo and others hold regarding the Sun.[23]

The statement might have been stronger in conclusion if Galileo's work had not been on the verge of condemnation by the Sacred Congregation and if Campanella had not been imprisoned when he wrote it. As it stands, however, the implications are there: he had been forced by new facts to adjust his philosophy in its central concepts.

Campanella may appear thus far to be remarkably advanced for his time in the history of thought because of his defense of Galileo and Telesio earlier, his emphasis on proof to the senses, and his acceptance of recent statements affecting the traditional structure of both matter and the universe, but this is not the complete story. His thought was not so clearly rational as this evidence implies; his was a truly paradoxical combination of new attitudes and old traditional notions. Those same walls that featured new knowledge in the *City of the Sun* were named for the seven planets and rose symbolically in ever-decreasing circles up a high hill to the temple on the crest. While "nothing rests on the altar [of the temple] but a huge celestial globe, upon which all the heavens are described, with a terrestrial globe beside it," these globes were often used by Solarians for very different purposes than might be surmised. Over the city floated the strange atmosphere created by astrologers who determined the proper time for important events, whether the planting of grain or the mating of human beings. Astrology combined weirdly with both science and religion in the *City of the Sun*.[24] The people "honor the sun and the stars as living things, as images of God," so while they worshiped only God, "they serve Him under the sign of the sun which is the symbol and visage of God from whom comes light and warmth and every other thing." Even the altar was shaped like the sun and the high priest was called the Sun. Long passages on astrology, usually deleted as unintelligible in English translations, were revealed in Professor Daniel Donno's new translation based on the 1602 Italian manuscript of the *City of the Sun*.

> When they founded their city, they set the fixed signs at the four corners of the world: the Sun in the ascendant in Leo, Jupiter in Leo preceding the Sun, Mercury and Venus in Cancer but so close as to produce satellite influence; Mars in the ninth house in Aries looking out with favorable aspect upon the ascendant and the *dator vitae* . . . Fortuna with the head of Medusa almost in the tenth house, from which circumstance they augur dominion, stability, and greatness for themselves.[25]

Since the city's establishment was so determined, it is natural to find odd practices retained in the enlightened society: bulls, of course, are bred when Taurus is in the correct position and sheep await the time under Aries. When the people pray, they face the horizon at one of the four points of the compass, "but the most intense and longest prayer is directed to the sky." It is possible that the misty, vague, religious unity with an astrological calendar controlling the daily life of these fortunate souls may simply have been the result of the Solarians' descent from a race in India that fled before the hordes of Tartars. But the roots of belief lie deeper in Campanella's background and, in this case, he did not apparently shift his position because of new knowledge discovered or opposing opinions held by respected contemporaries. He not only wrote six volumes with a supplemental seventh on the subject of astrology,[26] but in his personal life he believed devoutly in heavenly influences. In the stars he read of his own death in 1639 and tried unsuccessfully to avert it with propitiatory rites; always he was convinced that his birth was attended by six planets in the ascendant, a beneficent sign which determined his destiny—to be the reformer of the world.[27] In the letter of 1611 in which he complimented Galileo's report, he added that he had predicted the recent advances in astronomy from the conjunction of 1603, and again in a letter of March 8, 1614, to Galileo he defended the influence of the stars.[28]

Campanella's astrological attitude was compounded from many sources and supported by other facets of his philosophy. First of all, his high regard for Pythagorean doctrine and its first spokesman, Philolaus, was conducive to his belief. A veil of mystery has shrouded many of their teachings about the earth bound together with nine other spheres in a beautiful and musical harmony, but from such fancies have often come great values in the history of science. They introduced the important idea of the earth as a sphere and its revolution about the sun, as well as the concept of mathematical expression for the relationships of all parts of the universe. From them Plato took his ideas for the *Timaeus*, which was the most influential of his works surviving through the medieval period. The influence of heavenly bodies on man's body even as the great system of correspondency—unlimited analogies between the macrocosm of state or world and the microcosm man—is inherent in this thought. Campanella reflects both in his *City of the Sun*. In the extensive wall exhibits, for example, he observes that each type of fish is shown together with its "correspondence to celestial and earthly things," as well as to the arts and to nature. So, too, the herbs and trees are explained in "their relation to the stars." The concept is pervasive in all Solarian thought: they believe "the sea is the sweat of the earth, liquefied by the sun and uniting earth and air, as blood unites the human body and spirit. The earth is a great beast and we live within it as worms live within us." In similar words he expressed the same thought in one of his best lyrics entitled "The Universe":

> The world's a living creature, whole and great,
> God's image, praising God whose type it is;[29]

and he continues, saying that we are imperfect worms that in its belly have our low estate. We are like the lice upon the body of the earth which "is a great animal."

Campanella's was an animistic universe, a sentient nature, as was Bruno's, Telesio's, and Francis Bacon's. Telesio viewed the plants as having crasser spirits than animals and relied heavily on spirits in explaining bodily functions; he further credited the notion of spontaneous generation (as Bacon did) and believed in the existence of occult qualities.[30] His division of power between the hot heavens and cold earth was also encouraging to astrological interpretation. Thus, from the natural philosopher, the anticipator of modern empiricism, Telesio, Campanella found endorsement for his astrology as well as his penchant toward the occult.

In his early youth in Cosenza, he had been initiated into the occult world and necromancy by a young Jew named Abraham and became convinced that "great mutations, announced by astrology for the end of the century," would provide the opportunity for his world reform.[31] During this same period, it will be remembered, he was studying intensively the famous theologians like Aquinas and Albertus Magnus. It is to them and other church scholars and to the Holy Scripture that he turned for supporting evidence in his books on astrology, just as in the *City of the Sun*, after explaining the Solarians' belief regarding the Ptolemaic and the Copernican systems of the universe, he followed the passage immediately with: "They believe that what Christ said about the signs from the stars, the sun, and the moon is true." He referred to Matthew 24.29–30 regarding the coming of the Son of Man:

> Immediately after the tribulation of those days shall the sun be darkened, and the moon shall not give her light, and the stars shall fall from heaven, and the powers of the heavens shall be shaken.
>
> And then shall appear the sign of the Son of man in heaven.

On such bases as this from the Scriptures, both Aquinas and Scotus and many members of the major church orders had given at least qualified approval to astrology,[32] and some had actively practiced it.

In the details of his astrological system, Campanella primarily followed Ptolemy, who had compiled "all the rules of the Chaldeans, Egyptians, Indians, Greeks, and Latins, based upon observations made through many centuries."[33] He differed with the Ptolemaic tenet, however, that great alterations were caused by eclipses; instead, he adopted Cardan's Arabic doctrine of conjunctions, which we have seen operating in his fortunate birth date. The means of influence from the stars upon inferiors, Campanella believed, was through their heat, light, motion, and aspect. He studied astrology carefully, distinguished between those who practiced it superstitiously and those who approached it "scientifically," and, more important, he held with Albertus and Aquinas that "the will was not subject to the stars directly." Only when "sense rules and not reason" can the stars without doubt conquer.[34] And this he affirms in the conclusion to the *City of the Sun*, permeated by astrological doctrine:

> Know this: that these people believe in the freedom of the will; and they say that if a man, after forty hours of torture, will not reveal what he has resolved to keep secret, then not even the stars working so far off can force him to do so. But because the stars gently induce transformations in the senses, those who adhere to the senses more than to reason are subject to them.[35]

Campanella refers to the power of his own will, which carried him through torture and gave positive proof of its efficacy over the stars even on the Tuesdays and Fridays, his unlucky days, on which he says his tortures always came. This statement emphasizing the dominant role of will and reason concludes the utopia in which astrological forces have apparently been more powerful, even to the point of determining the pattern of life for the inhabitants.

This hybrid combination of astrology, the occult, and rational thought was typical of many men in the period of transition when they were searching for new philosophies while their roots clung to the past. Childish notions mingled indiscriminately with wisdom, and intuition often wildly accompanied scientific fact. Campanella's mixture of beliefs did not at all preclude his position in the vanguard of modern science. Such ideas were neither antagonistic nor mutually exclusive in the early 17th century or before. Kepler could report scientifically on the elliptical motion of the planets and, at the same moment, worship fervently and mystically the powers of the sun, and hold onto an essentially Platonic and Pythagorean idea of the universe. He was not skeptical regarding the claims of astrology; he looked for verification of the effects of heavenly bodies in the events of his own life.[36] So, too, with Tycho Brahe whose remarkable, systematic collection of astronomical data and the suggestion that a celestial body's movement might "not be exactly circular but something oblong" enabled Kepler to form his laws. It was Tycho who observed the upsetting new star of 1572, and in 1577 proved that a comet was above the moon's sphere,[37] both of which facts were devastating to the old Aristotelian universe in which there was a *fixed* number of stars, and changes only occurred nearer to grosser matter, that is only in the sublunary area. Yet Tycho could take time to make many astrological prognostications, dabble in alchemy, believe in "relations of occult sympathy between the 'ethereal and elementary' worlds, and entertain his mind with teachings of Hermes Trismegistus, Albertus Magnus, Paracelsus, and others."[38] Similarly, William Harvey and Sir Thomas Browne could take part in the examination of alleged witches and still vigorously support the new philosophy. No final, divisive lines had yet been drawn to distinguish science from magic; in magic lay the primitive man's attempt to explain and control natural phenomena, and his black powers have not been without occasional verification from scientific procedure: the witch doctor who used the mold from which penicillin was developed may also have performed miraculous cures. Of course, magic had to vanish finally into scientific method, but it did not happen without a struggle on the part of many minds. Even today we have among us those who hold fearfully to their superstitions and who follow closely their horoscopes. So, too, Campanella pictured his own philosophy with its contradictions in the perfect society he imagined, which was, nevertheless, the first utopia to give a leading role to natural sciences and base its structure on a scientific foundation.[39]

Johann Valentin Andreae

Johann Andreae's *Reipublicae Christianopolitanae Descriptio* appeared in print 17 years after Campanella's first manuscript of the *Città del Sole* in 1602, and his description of the happy land is less darkened by philosophical clouds; his people are exceptionally free from astrological influence and mysticism generally. No mention is made of magical practices or ceremonies bordering on the pagan sacrifice as was the custom of Solarians. Andreae's citizens take care to prevent the entrance of quacks such as "drug-mixers who ruin the science of chemistry." Neither astrology nor alchemy has a place in the society of Christianopolis. Repetition in two other passages emphasizes the point. When discussing metals and minerals, Andreae says:

> Here one may welcome and listen to true and genuine chemistry, free and active; whereas in other places false chemistry steals upon and imposes on one behind one's back. For true chemistry is accustomed to examine the work, to assist with all sorts of tests, and to make use of experiments. Or, to be brief, here is practical science.[40]

Again deliberately, in another laboratory "dedicated to chemical science and fitted out with most ingenious ovens and with contrivances for uniting and dissolving substances," Andreae is reassuring: "No one here need fear because of the mockery, falseness, or falsehoods of imposters; but let one imagine a most careful attendant of nature."[41]

He almost protests too much on this issue, and for the good reason that immediately preceding the publication of his utopia, Andreae had been closely associated with the Rosicrucians, a highly secretive order espousing a curious, unintelligible composite of beliefs in alchemy, the cabala, magic, and mysticism. They maintained relationships with both the Free Masons and the craft guilds. Rosicrucians enveloped all in the name of Protestant Christianity with the goal of world reform and the main target of the Anti-Christ, by which the pope was designated. Such spiritual movements were exceedingly popular in the early 17th century; some 200 works on this sort of subject appeared by 1624.[42] Ferdinand Maack, who writes most lucidly on this vague subject, describes the Rosicrucians' progress from Germany to England where it attracted men like John Dee, Robert Fludd, and Thomas Vaughan, twin brother of the poet. Between the years 1614 and 1616 four major tracts on Rosicrucian doctrine and history—the manifestos of the society—were published although several were written years earlier: the *Fama Fraternitatis*, *Confessio Fraternitatis*, *Die Chymische Hochzeit*, and *Allgemeine Reformation der gantzen Welt.* Andreae's authorship of the four works is generally accepted though it is difficult, except in the case of the *Chymische Hochzeit*, to prove finally because they were issued anonymously. The question of authorship is further complicated by the fact that at least two of them were finished much earlier, the *Chymische Hochzeit* in 1602 and the *Fama* was circulating in manuscript form by 1610,[43] which would mean Andreae was a very young man to have composed such works. He would have been only 16 or 17 at the time of writing the *Chymische Hochzeit.* This work, translated by Foxcroft in 1690, bore the English title: *The Hermetick Romance: or the Chymical Wedding.*[44] It is an extremely abstruse book, certainly not understandable to many, in which an allegory under the guise of a wedding is presented. The seven days of the wedding apparently correspond to developmental stages in the philosopher's stone, and these stages in turn are paralleled by references to Jesus Christ. The most recent commentators on the work, the educator Rudolf Steiner and Walter Weber, who has written an accompanying interpretive essay, consider the *Chymische Hochzeit* an intuitive work, revelationary in character and so possible for a young man to record. They think it can be appreciated only by readers spiritually perceptive. In effect, Steiner is the leader who, in our time, has awakened the memory of the movement that Weber feels was engulfed and destroyed by "the catastrophic philosophy of René Descartes, which has remained until today the undisputed foundation of agnosticism."[45]

Refusing to engage in battle with this spiritual, bodiless opponent, we attempt to decipher only Andreae's connections and position on the issue. He did accept authorship of the *Chymische Hochzeit*, referring to it later in his autobiography in which he explained that he wrote it as a game to show the folly of such belief.[46] Though some critics think he meant this statement that he was never serious, others feel that he was and said this simply to protect himself. The Rosicrucian movement was naturally suspect and scarcely a healthy, much less profitable group, with which to associate. So Maack, Steiner, and Weber all endorse the seriousness of Andreae's belief in the movement regardless of later denial. His other writings and actual efforts bear at least a similar, general attitude toward a Christian reform that was to be both divine and human.

Neither the *Fama* nor the *Confessio* is mentioned in the autobiography, probably for the same diplomatic reason, but Johann Arndt, the Lutheran leader and mystic to whom Andreae dedicated his *Christianopolis*, said Andreae told him in confidence that he and 30 others had joined in publishing the *Fama* as a trial balloon to see if there were secret admirers of "the real wisdom" and where they were.[47] The attempt was related to Andreae's effort in 1620 to organize a Christian society.[48] There seems little doubt of Andreae's authorship of the *Fama* and *Confessio;* in style and occasionally content they read like several of his other writings. And in the case of the *General Reformation* we need not be too concerned, because it is a translation of a chapter from Trajano Boccalini's *Ragguagli di Parnaso* printed in Venice in 1612.[49] Andreae's skill as a translator was notable and he was a constant student of languages, traveling throughout Europe to perfect them.

The content of the *General Reformation* is indicative of Rosicrucian interests and presents a program consonant with their aims. Boccalini had pretended to report from the realm of Parnassus where Apollo and wise men naturally ruled, but the bad state of the human world was cause for their concern. Since man was obviously deathly ill, had left the path of virtue, and was getting advancement in all the wrong and crooked ways, a general meeting was called to analyze reasons for his plight and recommend solutions such as equality of riches and goods. Various philosophers presented different solutions. Thales, in a group of seven Greek wise men, suggested that the world would always be imperfect and so it was best to leave it in its present condition. One wonders if Boccalini knew that Thales, the Ionian father of science, renowned for his mathematical enunciation of natural laws, had really recommended reform: a federal system for the cities of Ionia.[50] In any case, the clever, satiric pen of Boccalini was effective and used to endorse Rosicrucian doctrines of reform. It is a utopia in purpose, based on Christian theosophical philosophy,[51] and reflective generally of Andreae's own opinions.

In the other two Rosicrucian tracts, the *Fama* and its companion piece, the *Confessio*, which elaborates the philosophic thesis of the *Fama* in 37 principles difficult to separate, there is more evidence of Andreae's personal hand and thinking, which reappears in his *Christianopolis*. Since these tracts are not available in English [52] and are so little known, detail in describing them is justified and needed. Furthermore, Dr. Felix Held, who first translated the *Christianopolis* into English and remains the English authority on this work, quotes repeatedly from the *Fama* [53] to verify Andreae's early interest in a scientific college and his position on alchemy, but he neither gives the full picture which somewhat qualifies his conclusion nor does he consider at all the Rosicrucian tincture in Andreae's works. Therefore, it seems necessary to delve more thoroughly into this aspect of Andreae's thought.

The all-embracing doctrine of Rosicrucianism and of Andreae as well is most evident in the *Fama*, which purports to relate the biography of Brother Christian Rosenkreuz, father of the movement. Intermingled with the life history are seen the desire to reform, the attitude toward knowledge and learning, and the effort to win people to the cause. The *Fama* introduces us quickly to its purpose:

In as much as the wise and kind God has lately poured His mercy and kindness over the human race, our understanding of the universe and His Son has constantly increased and we have good reason to boast of a fortunate time. Not only has He uncovered for us half of the unknown and obscure worlds, but He has also given us many strange creatures and occurrences that have never before happened. He has also let enlightened spirits arise who are to restore the somewhat maligned and imperfect art of alchemy to its rightful place, so that man will finally discover his dignity and glory and recognize how he is but a microcosm and how deep his art is rooted in nature.

The scholars, however, do not want to come to an understanding. . . . Instead of collecting and sorting what God so generously has revealed to our century through the book of nature or the realm of the arts, one group slanders the other. People prefer, therefore, to stick to the old theories and to defend the Pope, Aristotle, Galen, and everything that resembles a codex against the shining, clear light, even though these authors doubtlessly—if they were still alive—would be happy to change their views. But we are too timid here for such a large undertaking and the old archfiend, despite the fact that the truth can be clearly seen in theology, physics, and mathematics, continues to display in heaps his cunning and his madness, and by using fanaticism and bickering hinders the right development.[54]

Thus the need for a general reformation is justified, and though Christianity with the Bible as The Book is the basic means constantly held before us, the advancement of learning is an integral part of the goal.

Brother Christian Rosenkreuz had started Greek and Latin by the age of five, learned Arabic in the cultural center of Damkar when 16, and early acquired his foundation for physical and mathematical knowledge as well as the occult sciences. Much impressed by the cooperation he saw among scholars in those far countries, he observes that

the Arabic and African scholars meet yearly and question each other about their arts, whether, maybe, something better has been discovered or whether practical experience has affected their calculations. And every year something is discovered that furthers the field of mathematics, physics, and the occult sciences.[55]

The author draws the moral for his own day in Germany where more could be accomplished if scholars, physicians, cabalists, magicians, and philosophers would only work together, if the majority of them would stop trying "to graze on the pasture alone." In the hope of converting European philosophers to new discoveries and teachings, Rosenkreuz went to Spain where he explained how the arts could be improved. He exhibited newly developed plants, fruits, and animals that did not follow the old laws of nature, and he gave them new fundamental principles underlying these changes. But the Spaniards laughed and were afraid to change their views; instead, they clung to their established knowledge. Discouraged, Rosenkreuz returned to Germany where he built himself a roomy and neat house, sat down to contemplate his travels and newly gained insights, and spent much of his time studying "mathematics and constructing many beautiful instruments for all phases of that art."[56]

After five years, he decided to gather together assistants and instigate the reformation. So the Brotherhood of Rosicrucians, a celibate group, was founded like a monastery in the new building called the Holy Spirit. The early Brothers included men from the professions of letters, art, and mathematics, a master builder, and a cabalist; their central task was to compile the magic language and dictionary which were still in use, according to the author, in the early 17th century. For the general public, however, the Brothers practiced medicine, healing the sick at no charge, so there were, of course, many patients in the cloister. Subsequently, it was decided the Brothers should go into different countries, extend their practice of medicine, check on their principles with other scholars to inform each other and to determine whether "observations in one or another country disclosed an error." Annually they were to return and report in detail on their activities.

The *Fama* continues with statements concerning the various

Brothers, their deaths and secret burials; focus is on the rediscovery of Father Christian Rosenkreuz' tomb[57] designed with triangles, circles, and squares in a most symbolic way to represent the universe. In the Father's hand was found the testament of the order that summarized his work, the result of "divine revelations, lofty imagination, and untiring efforts," which culminated in the creation of a "Little World," resembling in everything the "Big" one, and of a "compendium of all past, present, and future events." Among the treasures of art and literature found in the sepulcher were the vocabulary of Paracelsus (parodied by Joseph Hall, it will be remembered) and the other writings from which the Brothers quoted daily.

In the last few paragraphs of the *Fama* all readers are exhorted to consider the message,[58] join the Brotherhood, communicate with the members, and be careful of books and figures published under the alchemical name. Soon the Brotherhood will publish a list of these fraudulent works, the product of the devil. Again the *Confessio* issues the same warning

> to remind everyone most diligently to discard, if not all, though most of the books of the pseudo-alchemists who consider it a joke and a pastime to either misuse the holy Trinity for worthless things or to deceive the people with whimsically strange figures, and dark and obscure discourses, and to cheat the simpletons out of their money. Many such books have appeared . . . so that it becomes more difficult to believe the truth, in as much as the latter appears to be simple, artless, and pure while the lies appear to be magnificent, stately, and imposing. . . .
>
> Shun and flee these books, you who are smart, and turn to us.[59]

And the reader is promised upon joining the fraternity a "clear and understandable" explanation of the mysteries. The *Confessio* merely enlarges on the materials in the *Fama*, emphasizes the importance of the Bible as the heart of the program, and pictures enticingly the goal of one world with no illness, with all knowledge past, present, and to come in one great book.

The four Rosicrucian publications attracted great attention and were considered the foundation of the movement; hence

Andreae has been called the founding father, which is certainly erroneous.[60] Yet his exact relationship to the society cannot finally be settled. Unless new evidence is unearthed, it remains a controversial question. There is, however, no compelling reason to doubt his authorship of these documents: they are in his general line of thought and personal activity as will be shown, though his subsequent works are comparatively free from the secretive, mysterious element.

Whether he repudiated the movement later, no one can say, for when he seems to deny its existence or cast derogatory remarks at it, one realizes that this would have been the natural way both to protect the society's secrecy and oneself from slander. The movement's reception was stormy with attacks and defense. Thus Andreae seems to take an equivocal position: after the first slap in his *Christianopolis* at drug-mixers, he goes on directly to label them "imposters who *falsely* call themselves Brothers of the Rosicrucians." The italics are mine and serve to point out the fact that he distinguishes once more between the cheats and the honest Rosicrucians who practice the true alchemy.[61] In his introduction to the utopia, which starts "Hail, Christian Reader," he comments on the reprehensible conditions of this world and on the various reactions to reform movements. When he discusses the Rosicrucian movement, he discounts it but a serious tone belies his words, and one recalls that he, too, is a theologian.

A certain FRATERNITY, in my opinion a joke, but according to theologians a serious matter, has brought forth evident proof of this very thing [reaction to reform]. As soon as it promised, instead of the taste of the curious public, the greatest and most unusual things, even those things which men generally want, it added also the exceptional hope of the correction of the present state of affairs, and even further, the imitation of the acts of Christ. What a confusion among men followed the report of this thing, what a conflict among the learned, what an unrest and commotion of imposters and swindlers, it is entirely needless to say. There is just this one thing which we would like to add, that there were some who in this blind terror

> wished to have their old, out-of-date, and falsified affairs entirely retained and defended with force. Some hastened to surrender the strength of their opinions. . . . Others even embraced this with a whole heart. While these people quarreled among themselves, and crowded the shops, they gave many others leisure to look into and judge these questions. . . . Moreover, I am prone to praise the judgment of a man of the most noble qualities . . . who, when he saw that men were undecided and for the most part deceived by the report of that BROTHERHOOD, answered, 'If these reforms seem proper, why do we not try them ourselves? Let us not wait for them to do it.'[62]

Andreae agrees with the unnamed judge: the pattern of Christ's life alone is sufficient; there is no need to "learn the way of salvation and emulate it, from some society (if there really is such a one)—hazy, omniscient only in the eyes of its own boastfulness, with a sewn shield for an emblem and marred with many foolish ceremonies." So, like Robert Burton, who ironically abandoned hope in the Rosy-Cross men and decided to compose his own poetical commonwealth, Andreae concludes that the Christian way will be found in his "new REPUBLIC which it seems best to call Christianopolis," built, as he admits, for himself so that he may exercise the dictatorship.

The major concept Andreae, as a young man, may have learned from the Rosicrucians was his belief in reform, which dominated most of his writings. In the *Invitatio Fraternitatis Christi* (1617) he advocated founding a Christian brotherhood.[63] His epic poem *Die Christenburg*, written sometime around 1615, told the story of a righteous people beleaguered by an army of the Anti-Christ and devil, led by Tyrannus, Sophista, and Hypocrita (the same archenemies declared by Campanella). After the Christians heeded a wise man's advice and joined together in humility, God intervened and destroyed the enemy in a cataclysmic holocaust: the elements howled, the wicked ones crashed into an abyss, the animals of the forest and beasts of the wilderness were deployed to wreak vengeance. When all was over and quiet descended once more, the Christian citizens thankfully sang the victorious hymn of gratitude to their Saviour.[64] In the *Christianae Societatis Imago* and *Christiani Amoris Dextera Porrecta,* dated 1620, he held forth again his design for reform; so little doubt remains concerning his consistent and serious intent.

Only in the full-scale utopia of *Christianopolis,* by far his best work, do we see the society outlined in all its aspects. Andreae sets forth in the good ship, Phantasy, upon the Academic Sea, and after shipwreck reaches his haven, a triangular island in such beauty that one might think "here the heavens and the earth had been married and were living together in everlasting peace." The main city of this lovely land is Caphar Salama (village of peace), which is laid out in a perfect square with four towers looking toward the four quarters of the earth. We might judge from this description at the beginning of the work that allegory will be strong in Andreae's utopia, but such is not the case.[65] The Platonic atmosphere of correspondence between the great and little worlds may exist, but it does not affect the citizens' lives or their active pursuit of scientific data.

To the reader immersed in traditional utopias, it is suddenly startling to go on the long tour through laboratories in the natural sciences, astronomy, metals and minerals, anatomy, a "drug-supply" house, a hall of physics, and an "excavated place" for mathematical instruments. Never before in utopias has one encountered such a labyrinth of laboratories with analogous departments of education. The city is also functionally zoned with areas for specific types of industry related to the resources and raw materials required. The entire physical plan of Caphar Salama is designed with laboratories, equipment, classrooms, and factories to further the new science and its application. Naturally an attitude of open, liberal questioning is typical of Christianopolitans.

Not only have they gone beyond Campanella's Solarians, but they are well ahead of those living in the real world in their acceptance of Copernicus' theory. The citizens of Andreae's imaginary society in 1619 have already proved to their own satisfaction "the gentle rotation of the heavens," and they no longer fear "falling away by the motion of the earth." Furthermore, when they interrogate the stranger before admitting him to their city, they do not stop with questioning his moral and ethical standards, his belief regarding the church and word of God, or his knowledge of history and languages: Andreae says they wanted to know "what progress I had made in the observation of the heavens and the earth, in the close examination of nature, in instruments of the arts." Before this time there had been no hurdles of scientific knowledge to leap before entering ideal worlds. Now such examinations are acceptable, even necessary, for admission to the state. And sometimes they are administered to determine a candidate's qualification for a job. Campanella's high ruler in the *City of the Sun* also had to undergo tests, before his election, on the laws and history of the earth and heavenly bodies, the names of inventors in science, knowledge of all mechanical arts, physical sciences, mathematics and, as might be expected, astrology.

The singularly advanced views of Christianopolitans in scientific matters are directly attributable to Andreae's lifelong interest in this area. From early childhood, when tendencies toward mechanics and mathematics emerged, he continued to study these subjects. The greatest impetus probably came at the University of Tübingen, where he studied six years under the famous mathematician, astronomer, and early Copernican supporter, Michael Mästlin, who gave him the privilege of using his mathematical instruments and library. During this period, Andreae met Johann Kepler[66] who also had been a pupil of Mästlin, and they entered into a correspondence that lasted until Kepler's death. Andreae's mathematical studies were sustained and, in 1614, he published a series of lectures, the *Collectaneorum Mathematicorum Decades XI*, which he had delivered to a group of friends. Here he reveals his knowledge and understanding of the field; it is largely a collection of information with sources given. His intention was, as he states in the preface, to make available in one volume what has existed separately and so provide a basis for others to build upon. The work combines pure mathematics, the first two "decades" or divisions of the book, with its applications presented in the other nine parts; the proportion in itself is significant. The emphasis falls on the practical, and Andreae realizes the implications of his decision: in the third decade on Statics, the branch of physical science relating at that time to weight and its mechanical effects and to the conditions of equilibrium, which he considers the other sister of geometry (arithmetic was the first), he mentions Archimedes whose contributions often resulted from practical needs and commends the study.

> Statics does us a kind service, now through the water scale, now through the common scale, now through the lever and the winches; through wheels which move upon their axles and again through wheels which do not; through ropes and through screws. Statics occupies a unique and always useful position but is not always appreciated.[67]

Galileo already knew this and had found, in the tradition of Archimedes, that practical experiments in mechanics when combined with mathematics led to laws; Andreae was on the same target but not with the genius' clear-sighted aim. He had taken lessons from a watchmaker and learned also from a goldsmith and cabinet-maker; so he had experience in the solution of practical problems. Beyond this, he simply sensed their relationship to scientific principles. He discussed the watchmaker's art (Decade VI), relating it to a preceding discussion of the gnomonic method of telling time by the sun dial, and indicated his appreciation of the value of such work. To him automatics is

> that art which in our times is almost a rival of the heavens, and a helpmate of the most useful works. When one considers the intimate relationship that exists between mechanics and the exact sciences, or the mere calculation of so many revolutions, or the striking usefulness which mathematics here imparts to human life, one hardly has a reason to exclude it [automatics or mechanics] from the liberal arts.[68]

Other decades reproduce his lectures on the laws of optics in which he explains the fundamentals of perspective and demonstrates them in tables showing such structures as an archway, a winding staircase, and an arrangement of columns. The discussion proceeds with architecture divided into its civilian uses and its military applications in fortresses, bastions, and bulwarks. One table accompanying civilian examples shows a blueprint of Tycho Brahe's famous observatory, Uraniborg, which was located on the island Hveen, near Copenhagen. Brahe had designed the building and equipped it with newly devised instruments; he had even added a printing shop for its publications.[69]

When Andreae as a teacher lectured on astronomy (Decade IV), he was exceedingly objective, setting forth the four world systems then under consideration: those of Copernicus, Roeslin (a mathematician and physician living at the close of the 16th and during the early years of the 17th centuries), Tycho Brahe, and of the astronomer Raimarus Ursus or Reymers Bär. Tycho's view—the planets revolve around the sun which turns about a fixed earth—was an important one held by many until later in the 17th century because it satisfied the belief in geocentricity; it, too, was indistinguishable mathematically from the Copernican, and it had the further advantage of being an exact representation of what the observer actually saw.[70] Ursus' theory can be dismissed, for he stole Tycho's, after working as his assistant, and published it first with only one essential modification: he admitted the diurnal rotation of the earth.[71] Andreae gave Ursus attention, but his fairness, it will be noticed, did not extend to the point of including the Ptolemaic system. Its absence bespeaks the author's opinion.

His series of talks beginning with plane geometry continued through arithmetic and algebra to solid geometry. Unfortunately he did not state the proofs for the theorems, but Carl Hüllemann, who studied carefully the Collection and other pedagogic writings of Andreae, assumes these were given orally, and notes that in the decade on plane geometry Andreae often explained the use of diagrams given. Among the tables in solid geometry are two interesting drawings by the artist Albrecht Dürer, who was also a very competent mathematician and who conducted valuable experiments in optics, perspective, and anatomy. His work, as well as Holbein's, was in Andreae's private art collection.

Considered generally, the lectures formed a textbook on mathematics, offering important and basic theorems with factual brevity; it was a creditable job for the time, especially so for its recognition of the vital lines between practical work in mechanics and theoretical mathematics. Andreae, however, was a product of his period and interested in notions now known to be absurdities. The five Pythagorean-Platonic solids that fascinated Kepler were, of course, properly exhibited in tables, but when he finished listing different methods of measurement (indicating the absence of uniform standards, which troubled Robert Burton and Bacon, too), Andreae succumbed to the Pythagorean game and started fitting man into the various figures. For example, Table 96 pictures a man, whose center is the navel, standing with arms outstretched horizontally, the whole enclosed in a square bisected by two diagonal lines. The accompanying text reads as follows:

> Since man is the most perfect of God's creations, it is only right that he has been endowed with particularly harmonic dimensions which differentiate him from the rest. From these dimensions which form the basis, as it were, follow the other dimensions, and God himself ordered Noah to build his ark according to the dimensions of the human body: 300 ells long, 50 ells wide, and 30 ells high with divisions similar to man's.[72]

In a similar frame of mind, he seriously presents a table of centuries and millenniums subdivided intricately and characterized either by prominent historical leaders like Alexander the Great, or important events such as the deluge and the destruction of Troy. Although Andreae's outline is in squares, the scheme harks back to the great Platonic year that even in the 20th century enticed William Butler Yeats to devise his complicated synthesis.

Today such wandering into the marshes of imaginative philosophy in the midst of a textbook on mathematics would obviously invalidate the work; not so in Andreae's day. His *Collectaneorum Mathematicorum Decades XI* was a reputable book containing much information on mathematics and indicating a real appreciation of scientific problems. In his correspondence with the young Brunswick princes he counseled one of them,

Rudolf August, on the importance of adding "the eye of mathematics" to his other studies and that promptly, for he was just the right age to start the study. Andreae sent him a translation and explanation of Euclid, advising him to begin with easy chapters. Later he significantly praised the same Prince for not "caring for *astrologorum naeniae*"[73] (incantations of astrologers). Andreae's last pedagogic works continued to stress the value of mathematical knowledge and he devoted personal effort in its behalf. As supervisor of the Tübingen theological seminary, he departed from custom and hired a professor of mathematics, although he had to obtain the salary from a contribution by a local sponsor.[74] Thus the unusual concern for mathematics, which will be seen in the utopian *Christianopolis*, is substantiated by his own lifelong interest in the subject.

About the time of writing his utopia, Andreae encountered Campanella's works brought into Germany by Tobias Adami[75] who had visited the imprisoned Italian friar in Naples in 1612. Subsequently, Adami became Campanella's publisher and brought out the first printed Latin edition of the *City of the Sun* in 1623. Andreae was most impressed with his Italian contemporary, translated several of his sonnets, which were published,[76] and dedicated two apologues to him in the *Mythologia Christiana*. Campanella appeared in Andreae's eyes as a brother-in-arms, "an untiring heroic fighter against the heathen Aristotle."[77] The suggestion, however, that Campanella's utopia, which Andreae could have read in manuscript form, is reflected in his *Christianopolis*, as several critics claim,[78] cannot be seriously entertained aside from, perhaps, the very idea to write a utopia; Andreae had produced too much utopian writing generally before this time to allow credence to the theory of influence, and the parallels between his utopia and his other works, earlier as well as later, are stronger than can ever be shown with Campanella's *City*.

Francis Bacon

Oddly, not the work of the free, Lutheran minister, Andreae, but the writings of the Dominican friar in prison reached Francis Bacon, if we accept only his own words as evidence. He mentioned Campanella just once and correctly placed him with Telesio and others. The reference was not complimentary; nor did he praise Telesio in this instance. Though he had placed him among the first of the new men, the Italian thinker was guilty of constructing prematurely a philosophical system from inadequate evidence. For Bacon, this was an age-old trap that had caught the minds of many men:

> In ancient times there were philosophical doctrines in plenty, doctrines of Pythagoras, Philolaus, . . . Democritus, Plato, Aristotle, Zeno, and others. All these invented systems of the universe, each according to his own fancy, like so many arguments of plays; . . . Nor in our age . . . has the practice entirely ceased, for Patricius, Telesius, Brunus, Severinus the Dane, Gilbert the Englishman, and Campanella have come upon the stage with fresh stories, neither honoured by approbation nor elegant in argument. . . . Each has his favourite fancy; pure and open light there is none; every one philosophises out of the cells of his own imagination, as out of Plato's cave.[79]

In an earlier work Bacon had already cautioned lest anyone should suppose he was "ambitious of founding any philosophical sect, like the ancient Greeks, or some moderns, as Telesius, Patricius, and Severinus."[80]

His deep distrust of unaided reason and fear of invasion by the sprite "fancy" account in part for his failure to appreciate the great discoveries of the Renaissance. He leveled the same musket again at Gilbert, on whom he was excessively hard. After commending him for "having employed himself most assiduously in the consideration of the magnet," Bacon condemned him because he "immediately established a system of philosophy to coincide with his favorite pursuit."[81] The system Gilbert presented in the last section of his book *De Magnete*, however, was certainly no figment of his imagination coinciding with his "favorite pursuit;" it was Tycho Brahe's system of the universe that Gilbert described although he did not name Brahe.[82] Elsewhere Gilbert did acknowledge his source—in the work *On Our Sublunary World, a New Philosophy*, which was not published until 1651, long after his death, but internal evidence shows it was completed by 1612. Bacon knew it in manuscript and used several passages from it in his *History of Winds*.[83] Still he was not content to let Gilbert's soul rest in peace; in the most derogatory of all comparisons, he discharged another volley on worse grounds: Gilbert was guilty of insufficient evidence, which was a common fault in the empiric school "confined in the obscurity of a few experiments." Alchemists and their dogmas were a strong example of this, and it was difficult to find another so bad "unless perhaps in the philosophy of Gilbert."[84]

The same general complaint was made regarding Galileo and Copernicus. Bacon recognized the "noble discoveries" of Galileo and gave a full summary of the facts reported in the *Starry Messenger*, finding them excellent "so far as we may safely trust to demonstrations of this kind; which I regard with suspicion chiefly because experiment stops with these few discoveries, and many other things equally worthy of investigation are not discovered by the same means."[85] Copernicus had simply lacked sufficient physical data; it was the same reason Campanella had given for the Solarian attitude toward the Copernican system. Both authors at least agreed on the importance of examining nature—the factual, concrete object—and proof to the senses.

In effect, Bacon was defending his own method in opposition to all other schools of approach whether ancient or modern. Even though Gilbert had used the inductive method generally and had a clear view of the place of observation and experiment,[86] it was not sufficiently convincing to Bacon. And Harvey, who was Bacon's personal physician, had also used this approach, but his work Bacon noticeably did not discuss.

Bacon's contacts with outstanding men of science were not personal and direct except, of course, with Harvey. If Bacon talked as he wrote of them, he certainly would not have been especially beloved in scientific company unless it were for his powerful prose expression. Not only did he undervalue his contemporaries' writings, he seems not to have known of some major discoveries. Kepler had dedicated his *Harmonice Mundi*, which contained the third of his planetary laws, to King James I when Bacon was Lord Chancellor, yet he does not mention Kepler.[87] He did not correspond with noted discoverers, as Campanella did with Galileo and Andreae with Kepler; yet his correspondence was extensive and through go-betweens he attempted to establish relationships, circulate his own writings, and learn the latest news of work abroad. Three copies of the *Novum Organum* went to Sir Henry Wotton, the English ambassador in Venice with wide acquaintance in Europe, who promised to give one to Johann Kepler.[88] Before this, Bacon had a rather tenuous connection, via his lifelong friend and informant from the Continent, Toby Matthew, with Galileo on the specific subject of the tides. In a letter dated April 4, 1619, Matthew reports on Galileo's answer to Bacon's discourse on tides. Matthew says a certain gentleman, Richard White, whom the letter will introduce and who is bringing to Bacon several of Galileo's works, had recently spent considerable time in Florence. It was he who furnished Matthew with the information now being forwarded to Bacon: namely, "that Galileo had answered your discourse concerning the flux and reflux of the sea, and was sending it unto me; but that Mr. White hindered him, because his answer was grounded upon a false supposition, namely, that there was in the ocean a full sea but once in twenty-four hours." "But now," Matthew assures the Lord Chancellor, "I will call upon Galileo again."[89] Besides the books Mr. White was delivering, there were two of Galileo's unprinted papers, one on the tides and the other on the mixture of metals.

Galileo's "answer" to Bacon was one of the unprinted papers, probably what White had criticized; it was a copy of a letter Galileo had written to Cardinal Orsini on the subject in 1616,[90] and two years later (just before White's visit) had revived and sent to the Austrian Archduke Leopold with a new preface attached describing it as "merely an ingenious speculation." Whether Galileo really thought this or, after being commanded to be silent on such issues by the Holy Office, he added the phrase as deliberately deceptive, one cannot know. The tract was an attempt to give physical proof of the Copernican system, his thesis erroneously being that tides were due basically to the earth's motion. In any case, Bacon took it seriously and, in the *Novum Organum* (2.46) published in 1620, wrote:

> It was upon this inequality of motions in point of velocity that Galileo built his theory of the flux and reflux of the sea; supposing that the earth revolved faster than the water could follow; and that the water therefore gathered in a heap and then fell down. . . . But this he devised upon an assumption which cannot be allowed, viz. that the earth moves; and also without being well informed as to the sexhorary motion of the tide.

Bacon's criticism here is perfectly just; the most interesting remark, however, is his refusal to grant the earth's motion. True to his distrust of theoretical systems, he is consistent, and in the utopia of *New Atlantis* there will be no conjecture on this point as there was in Campanella's *City*, and certainly no acceptance as in Andreae's *Christianopolis.*

A brilliantly clear and pure light illuminates this ideal kingdom as Bacon would have it; shadowy regions are completely dispelled. In the *New Atlantis* there is no trace of Campanella's astrology, no evidence of Andreae's problem in differentiating the true from false alchemy. Bacon had spoken plainly, and several times, on these issues elsewhere:

> The sciences themselves which have had better intelligence and confederacy with the imagination of man than with his reason, are three in number: Astrology, Natural Magic, and Alchemy; of which sciences nevertheless the ends or pretences are noble. For astrology pretendeth to discover that correspondence or concatenation which is between the superior globe and the inferior: natural magic pretendeth to call and reduce natural philosophy from variety of speculations to the magnitude of works: and alchemy pretendeth to make separation of all the unlike parts of bodies which in mixtures of nature are incorporate. But the derivations and prosecutions to these ends, both in the theories and in the practices, are full of error and vanity; which the great professors themselves have sought to veil over and conceal by enigmatical writings. . . . And yet surely to alchemy this right is due, that it may be

> compared to the husbandman whereof Aesop makes the fable, that when he died told his sons that he had left unto them gold buried under ground in his vineyard; and they digged over all the ground, and gold they found none, but by reason of their stirring and digging . . . they had a great vintage the year following: so assuredly the search and stir to make gold hath brought to light a great number of good and fruitful inventions and experiments, as well for the disclosing of nature as for the use of man's life.[91]

Again, in the *Novum Organum* (1.74), he reiterates the opinion that though good may come from alchemistic endeavors or natural magic, it is by chance, and few discoveries of import have resulted. His is an intelligent disposition of the issue and, thinking thus, he omits these questionable areas in *New Atlantis*. He retained these views on alchemy and natural magic, but revised his opinion of astrology in a later publication, *De Augmentis* (1623). Here he advocated a Sane Astrology, cut out the greater fictions from the study, and recommended four means to purify it. Once this was done, he would trust it in predictions and, with additional cautions, in elections. Thus he did not categorically reject astrology; he emerged as a moderate astrologer and his attitudes were reminiscent of others before him as well as contemporaries.[92]

More peremptorily he dismissed Pythagorean thought as superstition, "coarse and overcharged," and he feared the Platonic development of it as even more dangerous because it was more refined.[93] Summarily, he banished "Fortune, the primum mobile, the planetary orbits, the element of fire, and the like fictions, which owe their birth to futile and false theories."[94] These are examples of his idols imposed upon the understanding. And he went further in another passage in the *Novum Organum* and again in the *Parasceve* to explain that he could not accept the traditional four elements of fire, air, earth, and water. They had been named with undue regard for the senses and their perception; the so-called elements were not first principles of things but larger masses of natural bodies.

Although he reported a couple of instances of divination, allowed a miracle to introduce Christianity into the Bensalemite society (miracles were consigned by many to the early days of Christianity when they were needed to establish the faith), and still held with an animistic universe, superstition and the occult were swept out: he maintained that bodies were not acted upon except by actual bodies.[95] No spirits, then, were found wandering about in *New Atlantis*, the light was too intense, unmasking all phenomena in clarity. "His Lordship's course was to make wonders plain, and not plain things wonders."[96]

Light, to Bacon, was truth, the highest knowledge, the first creation by God, and consequently it was foremost in his imagery in *New Atlantis* even as it was in his essay *Of Truth* and many other writings. In the utopia, Bacon's Merchants of *Light* seek knowledge from all parts of the world; *Lamps* sift it out and direct "new experiments of a higher light;" and Salomon's House, the scientific center and outstanding feature of the society, is not only the "noblest foundation that ever was upon the earth," but the very *lantern* of the kingdom. No doubts exist: from the light of knowledge—scientific—come the fruits that citizens of Bensalem enjoy.

Bacon's utopia was a complete testimony of his belief that through scientific knowledge man may progress to the utopian world. And from all his works on the new method of learning and its promise for mankind, the *New Atlantis* emerged as the lively embodiment of his theory, his ideas in operation. When William Rawley, his close friend, first presented the work to the public a year after Bacon's death, he explained in his letter to the reader that Bacon's fable was devised "to the end that he might exhibit therein a model or description of a college instituted for the interpreting of nature and the producing of great and marvellous works for the benefit of men." It was to display his scientific method applied, to demonstrate his own favorite idea, and to describe as reality the results he so optimistically expected. At the heart of the society was the institution, Salomon's House, famous for its great wisdom, or the College of the Six Days Works as it was also called, because its purpose was to discover the secrets of those days of creation. As the laws of the universe were understood by man, it was confidently expected that he would in turn create a new and better world—a New Atlantis. The various houses for experimental work with their ingenious instruments and equipment were already producing amazing results. This was the new way and the only way to the perfect world that man had the power to form.

It was to this belief that Bacon—regardless of how he evaluated his contemporaries' efforts—dedicated all his writings. The man who wrote of science like a Lord Chancellor, as Harvey wryly commented, was an effective "bell-ringer" who called "the wits together" for one purpose: to propagate the new thought. Bacon saw himself in this role as Homer's trumpeter who summoned his fellows to activity, not for contention but for unity, so they could storm and capture the heights and strongholds of nature and extend the boundaries of human empire. No man of the period wrote more on the advancement of learning, the reorganization of knowledge, the need for investigation conducted without preconceived notions, and the standards of proof necessary before evidence became truth. His total campaign was carefully designed, and his major writings were assigned exact positions in the over-all plan for this *Instauratio Magna.* He felt sincerely that he personally should start the task of collecting information and of asking the key questions to provoke others to join the effort. Typically, his works end with a list of questions to be answered or categories to be explored; typically too, they are unfinished, as was the *New Atlantis*, because he felt he should get at the basic, brick-layer's job and start the new edifice. Bacon's earnestness can never be suspected; even his death resulted from the unhappy decision to leave his coach on a February day to try stuffing a chicken with snow to see if the cold—refrigeration—might preserve the meat.

In this devotion to the cause the new utopists joined the Lord Chancellor: each worked practically and publicly to effect his reforms. Their utopias were a means to that end, but none stopped with writing. Campanella's *City* was directly related to his attempt for a reformed society in Calabria. He wrote it in the vernacular (while philosophic works were in Latin for the learned audience) to gain adherents, and the title page of the first manuscript says that in this work "is demonstrated the idea of reform of the Christian Republic according to the promise God made."[97] From his earliest plans, in which he thought a grand celestial conjunction would produce the universal kingdom, to the time, four years before his death, when he proposed a plan to Richelieu for a New Christianity in which all the world would unite under the banner of France, he continued to nurture schemes for perfection. Even Columbus' discovery of the New World was part of God's plan to gather the whole world together under one law.

Neither did Andreae stop with paper plans for reform; he personally sought to bring the dream of *Christianopolis* into existence. He had tried various Christian fraternities, but about 1620, a year after the publication of his utopia, he made an attempt that succeeded. Starting with his own congregation in the church at Calw on the Nagold, he set up an ideal social system that soon influenced others outside his church. They formed a mutual protective association for workers in the clothing and dyeing industries that continued into this century as a well-endowed organization.[98] That Andreae's efforts should culminate in a workers' union is especially interesting in view of his concern for the laborer in his utopian society. He was unique among utopists of the earlier part of the century in the attention he paid to unskilled workers and to trades and industry generally. Though religiously oriented, his belief in the new science and his interest in the practical aspects of life come forth as the major means to effect the spiritual reformation he envisaged.

Bacon worked less obviously, less directly, in a sense, for his reform. He did not seek individual followers as Campanella and Andreae attempted to recruit them. Only in an early, private memorandum dated July 1608, do we find Bacon considering the possibility of actually establishing a college for research. His plan read like brief notes for the *New Atlantis*, and he wished to obtain a post at Westminster, Eton, Winchester, or preferably Trinity College at Cambridge, where he could command both "wits and pens" in compiling two histories of interest to him at the time. One was to be a history of marvels—that is, nature erring from her usual course, the other a history of observations and experiments in the mechanical arts.[99] The plan was never inaugurated, and in subsequent efforts he leaned more heavily on his powers of persuasion in writing, which were not the least in vain.

The three new utopists were dedicated reformers whose minds were challenged by scientific development and they, in turn, would further it, spread the word, and show its promise in utopia. Whether the imaginary country lay near the equator like Andreae's, in Taprobane in the Far East where Campanella's City was located, or in the South Seas where the New Atlantis arose, it was an ideal model for their own countrymen to work toward. Furthermore, it was contemporaneous—not the Golden Age of the past or the Millennium to come. Their utopias could, by man's diligence and industry, become reality in the immediate tomorrow.

Primarily as publicists they served science, or more accurately, the scientific attitude, for it was to the general understanding and acceptance that they contributed; as middlemen between the scientist and the public, they translated the effects of research into meaning for man's life; and as salesmen, they effectively encouraged the public to buy the new, the latest doctrine, and to join the attempt. Such intelligencers, as they would have been called in the early 17th century, were a powerful force operating in behalf of the new knowledge caught, as it was, between man's fear of the novel and strong adherence to past belief.

More important, the new utopists breathed an optimistic spirit, infusing it colorfully into utopia. To each of them the vision was

> not of an ideal world released from the natural conditions to which ours is subject, but of our own world as it might be made if we did our duty by it; of a state of things which . . . would one day be actually seen upon this earth such as it is by men such as we are; and the coming of which he believed that his own labours were sensibly hastening.[100]

Andreae, Campanella, and Bacon earnestly believed their "labours were sensibly hastening" the advent of the ideal society that this world would soon see. No longer do we hear the utopist say with Plato that he has no hope of realizing the actual foundation of the *Republic;* nor does he lament with Sir Thomas More that *Utopia* is rather to be wished for than hoped for. With remarkable energy and optimism, the new utopists worked hard for the adoption of their plans. The basis of their enthusiasm, the source of their great energy, was the new learning. With scientific knowledge man was to build the road to progress and the better world.

V

THE DOCTRINE SPREADS

THE UTOPIAN CRY OF BELIEF in the new learning so eloquently uttered by Andreae, Campanella, and Bacon echoed first and most widely in England, where reform movements were rampant in the 17th century and where Bacon's persuasive, powerful voice could, of course, most easily be heard. Andreae had no such influence in Germany, which was torn by the Thirty Years' War; he was known locally, and his earlier works were published anonymously. Similarly, there was no great reform movement following Campanella's plans; the Catholic Holy Office effectively checked dangerous developments and held tight, restrictive controls for some time longer. Yet both Campanella's and Andreae's works were known in England.

Here the Civil War and deposition, even beheading, of a king, failed to deter utopian dreams; instead it seemed to spawn them. Unrest, reform, and revolution went together and did not stop with effects in just one area of the social culture: religious sects, each believing in its own infallibility, and hence at war with the established church, political dissension over the king's rights and those of his commoners, economic upheaval accompanying industrial growth and extension of land ownership, emotional and mental disturbance caused by the discovery of additional and critical knowledge, geographic as well as scientific—all were occurring simultaneously in the 17th-century Englishman's life. Joseph Glanvill saw some advantage in such an "unhappy Season," since young academicians "were stirr'd up by the general Fermentation that was then in Mens Thoughts, and the vast variety that was in their Opinions, to a great activity in the search of sober Principles, and Rules of Life."[1] The atmosphere bred a rash of utopian writing, each tract pointing out, as the author saw it, the desirable path for society in order to correct its present ills. Men's thoughts were obviously stimulated, and Sir Thomas More's proper noun "Utopia" created for his particular society became, in effect, a common noun.[2]

Another reason for the popularity of the utopian genre at this time may well have been its protective coloring: the fictional element gave the author freedom to express ideas that might otherwise have been suspect. It released him from direct responsibility for his statements. The utopist could say, and often did, that the account given did not necessarily represent his personal views, that he was merely repeating what had been told to him by his thoroughly "reliable" informer, while of course the provocative mischief was done: the ideas were loose in circulation. This was an especially attractive feature of utopias in a century when censorship prevailed and ears or head might be lopped off for small cause. The fate of Giordano Bruno, Galileo, and Campanella was not easily forgotten. And in Protestant England, although treatment generally was less severe, outstanding men like Henry Oldenburg, founder of the *Philosophical Transactions* in 1665, were occasionally imprisoned "for dangerous designs and practices." Oldenburg's heavy foreign correspondence as Secretary of the Royal Society evidently caused investigation, for Samuel Pepys recorded in his *Diary* that he had heard Oldenburg was in the Tower "for writing newes to a virtuoso in France, with whom he constantly corresponds in philosophical matters; which makes it very unsafe at this time to write, or almost do anything."[3] The utopian format offered at least a relative safety to the author in such perilous times.

Then, as now, the excitement of reform was contagious and spread rapidly, enticing men from many and varied walks of life and creating bonds of friendship among men who, aside from their interest in reform, seem often to have had little in common. Extravagance characterized utopian proposals and optimistic hopes alike; it had been shown that not even the sky was a limit, nor were the "Pillars of Hercules" in the Strait of Gibraltar any longer the boundary of the known world. Bacon's motto had been *Plus Ultra* and some men were beginning to believe it. For presenting ideas of the "more beyond" there was no more natural literary genre than the utopia. This was its basic purpose.

To many, Bacon's *New Atlantis* was the perfect type in both form and content. Although Bacon's influence reached a peak in the decades immediately before and after the Restoration (1650–1670), the imprint of his ideas as expressed in Salomon's House was apparent earlier in utopias. His college of light was clearly reflected not only in numerous imaginary colleges but also, according to Glanvill and many others, it was the *Romantick Model* for the Royal Society, established later. The line of development leading to the Society's foundation was, in fact, closely related to the utopian authors of scientific bent. Their plans varied from Bacon's in detail, and frequently they were narrower in scope. Instead of describing the society completely, authors often presented partial utopias—utopian tracts—devoted to one area of development like the scientific college, or they concentrated on applications of the new information in specific fields like husbandry and medicine. Nevertheless, recalling the Bensalemites in *New Atlantis*, citizens of the utopias that followed were convinced that the better life was to come from scientific research. And the sundry houses for experimental research with their clever devices that had already led to great discoveries, as reported by Bacon, were reproduced in subsequent utopias, sometimes as almost exact replicas and, in other instances, as the basis for expanded functions.

Samuel Hartlib and His Group

Foremost among the group of disseminators spreading the Baconian type of doctrine shortly after the Lord Chancellor's death in 1626 was Samuel Hartlib, a Polish émigré to England, who was an indefatigable agent for reform. He must have been a desperately busy man, for he apparently lived in a state of constant correspondence with many men more important than himself while his home functioned as a valuable center for discussion and the hatching of new schemes. The papers he left were a voluminous mess of countless ideas and notions promoted for education, husbandry, religion, and science; it is indeed often difficult to determine which were his and which were sent to him by congenial friends thinking along the same lines.[4] One is typical: *Cornu Copia; a Miscellaneum of Luciferous and most Fructiferous Experiments, Observations, and Discoveries immethodically distributed* . . .[5] Yet he was most effective as a communicator and catalyst. One of his friends called him "a conduit pipe for things innumerable," and said he had set more men to work both within and without Great Britain than perhaps anyone of his rank and position ever did.[6] Undoubtedly Hartlib asked stimulating questions, served to inform others of what was going on, and encouraged those who were more talented than he. John Milton wrote the tract *On Education* at his request and William Petty, a younger acquaintance of Hartlib's, framed his proposal for the *Advancement of some particular Parts of Learning* for Mr. Samuel Hartlib.

Into Hartlib's more or less scientific circle came several well-known foreigners who had migrated to London, bringing with them important connections with men on the Continent working along similar lines. One was Theodore Haak, a German refugee and one of the main correspondents in England with the French friar Mersenne,[7] who provided a center for men of scientific interests in Paris at his convent on the Place Royale, similar to Hartlib's headquarters in London. Haak not only helped build the vital communications system with the Continent, but he became an original Fellow of the Royal Society in England. He was, in fact, credited with having first suggested in 1645 the regular group meetings of the "Invisible College," a forerunner to the establishment of the Royal Society.[8]

It was because of Hartlib's urgent invitation in which members of Parliament concurred that Jan Amos Komensky (Comenius), the prominent Czech educator, arrived in England in 1641 to present his plans for a great and extensive project, an all-embracing *Pansophia*, an encyclopedic effort in which universal knowledge was to be gathered. Comenius had not only studied Bacon's thought carefully and linked him with Campanella as the "glorious restorers of Philosophy,"[9] but he also had read Andreae's works, which he saw fanning the sparks of the new learning into a flame. Andreae's were "golden writings" to Comenius, who lauded him in the introduction to the Czech edition of the *Didactica Magna:* "He should precede everybody because of his importance; since 1617 he has, in a number of writings clearly shown what ails the church, the state, and the schools, as well as indicated the desired remedies for their illnesses."[10] The two men were in direct correspondence and, as early as 1629, Andreae had written his plans for founding a Christian fraternity to Comenius.[11] Another of Hartlib's friends, the scientist Robert Boyle, wrote to him in 1647, asking that a translation be made of Campanella's *Civitas Solis* and of Andreae's *Christianopolitana,* as he referred to them.[12] In the same year Samuel Hartlib engaged Jeremy Collier to translate Campanella's utopia into English (the manuscript has not been found).[13]

Although Andreae's and Campanella's works were known and requested, it was naturally Bacon's doctrine that spread most prolifically in England and, in the 1640's, especially through the efforts of Hartlib's group of reformers. Many of those espousing Bacon's works, however, were not fulfilling his request or carrying out his plans for collecting data; rather it was the Baconian spirit and general inductive approach (not the Baconian method itself) that inspired the group's endeavors and encouraged each individual to create an original scheme for improvement. So, for example, Comenius recognized Bacon's contribution but saw his own going further:

> Such a standard [to separate truth from falsehood] the illustrious Verulam seems to have discovered for scrutinising Nature, a certain ingenious induction, which is in truth an open road by which to penetrate into the hidden things of Nature. But this induction demands the incessant industry of many men and ages, and seems so laborious, and keeps certainty so long in suspense, that notable as the invention is, it is contemned by many as useless. And indeed it brings small help to me in building Pansophia, because (as I have said) it is ad-

dressed solely to the revelation of Nature's Secrets, whilst I look to the whole Scheme of Things.[14]

Despite his criticism, Comenius' *Great Didactic* was similar to Bacon's *Instauratio Magna* in general purpose: both men looked for state support to make possible the new revelations in knowledge, which were to be derived from observation of natural phenomena, and both believed their method universally applicable. While Comenius was less entranced with the scientific laboratory and projected his pansophic theories more into the educational process than Bacon, they were brothers in effort, and Comenius, as an impressive foreigner in London, played no little part in creating the Baconian legend.

One odd item of curious note enters with Comenius: this traveling teacher, abounding in new philosophic theories, possessed from 1614 until 1641, when he visited England, the only manuscript copy of Copernicus' *De Revolutionibus* that has come down to us.[15] Fortunately, it was not in the fire in which Comenius lost many of his papers, including his refutations of the Copernicans and Cartesians.[16] But at heart Comenius' interest lay in the application of natural philosophy to the educational process, which led him into such slippery, absurd analogies as to claim that since "the sun is not occupied with individual objects, tree or animal, but lights and heats the whole earth," so in imitation of this "there should be but one teacher at the head of a school or, at least, of a class." He further observed that nature progressed from the easy to the more difficult; therefore children in the classroom should start out with the simple and known, then move gradually to more remote, abstract, and complicated knowledge.[17] Many of his educational theories are today practiced, but little assistance to science could come from such an approach. His greatest service here was complete endorsement of new ways to knowledge and consequently the abrupt dismissal of stifling past authority. He held for an inductive building process that he, like Bacon, believed all men capable of performing.

Nevertheless, ideas for establishing a corporation of learned men in England, which had brought Comenius over with such high hopes he thought "nothing seemed more certain than that the plan of the great Verulam respecting the opening somewhere of a Universal College wholly devoted to the advancement of the sciences could be carried out,"[18] did not materialize. The Universal College, which was to be located in England as a memorial to Bacon, failed to gain support of the government, occupied at that time with an incipient civil war which broke out the following year (1642), and Comenius left London. En route to his next post in Sweden, he stopped at Endegeest near Leyden for a four-hour visit with Descartes during which Comenius maintained, in opposition to the Cartesian theory, that all human knowledge derived from the senses alone and reasonings thereon were imperfect and defective. Comenius reported that, though their positions were poles apart, they "parted in friendly fashion."[19]

Comenius' leaving in no way discouraged the ebullient Hartlib and those in constant communication with him; he had published his own utopia in the year of the Czech educator's visit and continued to work devotedly toward its accomplishment all his life. To the famous Robert Boyle, formulator of the law of gases that bears his name, Hartlib wrote concerning his utopian *Description of the famous Kingdom of Macaria*,[20] that it was "most professedly to propagate religion and to endeavor the reformation of the whole world." Like Andreae, two of whose utopian tracts for a Christian society were found among Hartlib's papers,[21] he felt no cleavage in combining his religious and scientific goals; on the contrary, the scientist was interested too in completing the job of the Reformation. Boyle remained strongly disposed toward religion, and Theodore Haak, while promoting scientific investigation on the one hand, continued his major work as a translator on the other, receiving from Parliament in 1648 the sole right of translation into English of *The Dutch Annotations on the Bible*.[22] Hence Hartlib's *Macaria* reads like the work of a religious missionary, but one who has adopted the new methods to attain his dream. He urges divines to preach this doctrine and answers the pessimist who feels that such reforms will not come "before the day of Judgment." Science is widely applied in his perfect society, which has erected a "college of experience where they deliver out yearly" the latest medical information and report on inventions for the general welfare of man. As a result, Macarians "live in great Prosperity, Health, and Happiness;" they have rightly ordered their affairs and sensibly established five "under-councils" to supervise activities in husbandry, fishing, trade by land and by sea, and to plan for the surplus population. To do the latter, one council manages new plantations for emigrants, a very important program when one remembers that North American and Caribbean colonies were then in the process of establishment under England's flag. Hartlib's is a practical society that copes with its economic problems, plans for the health and welfare of its citizens, and expects utilitarian benefits constantly from the scientifically geared college of experience that meant to Hartlib "experimentation."

He fully recognizes his utopian predecessors in the remarks of his scholar discussing the famous kingdom of Macaria with a traveler who has been there:

> I have read over Sir Thomas More's Utopia, and my Lord Bacon's New Atlantis, which he called so, in imitation of Plato's old one; but none giveth me satisfaction, how the kingdom of England may be happy, so much as this discourse, which is brief and pithy, and easy to be effected, if all men be willing.[23]

With this definite purpose, Hartlib dedicates the utopian tract to Parliament, confident "that this honourable court will lay the corner-stone of the world's happiness, before the final recess thereof." He wants action in the current session of Parliament for England's immediate benefit.

For years Hartlib had promoted such reform: as early as 1627 he attempted to found the Antilian Society which, G. H. Turnbull believes, furnished the pattern for Macaria and may have been directly related to Andreae's efforts during the same period in Germany.[24] The geographical locations most favored by Hartlib for the new colony appear from his correspondence to have been Virginia in the New World and an island off Liffland in Poland. Like the majority of Hartlib's plans, however, this one too ended unsuccessfully, even though he persisted throughout his entire life in attempts to found Antilia. In 1660 the scheme was revived when a Swedish nobleman, Lord Skytte, promised support for it, but in 1661, just a year or so before Hartlib's death, he sadly wrote his friend Dr. John Worthington: "Of the Antilian Society the smoke is over, but the fire is not altogether extinct. It may be it will flame in due time, though not in Europe."

With Comenius he labored to feed the flame of free thought and, although Hartlib scientifically got no further than turning his kitchen into a laboratory—from which no remarkable discoveries issued forth—he supported, encouraged, and financially assisted others. He was never a member of the Invisible College or the Philosophical Society, which led to the Royal Society. Yet he fostered the development of the younger William Petty and knew Christopher Wren, Abraham Cowley, John Evelyn, Robert Boyle, and many others who were early Fellows of the Royal Society.

On behalf of Petty (1623–1687) he energetically worked to obtain an appointment at Gresham College for the advancement of experimental science and mechanical knowledge. Petty was not chosen for the post, but he became a valuable participant in the scientific movement, largely on the practical and applied side, and he revealed in his writings many lessons learned from Hartlib. His utopian proposal, *The Advice of W. P. to Mr. Samuel Hartlib, for the Advancement of some particular Parts of Learning,*[25] was printed in 1648, the year he went to Oxford and joined Boyle, John Wilkins, and others in the Philosophical Society. They met in Petty's living quarters conveniently located in the home of an apothecary, whose resources we may be reasonably sure did not lie unused. Experimentation was in all directions, but Petty's interest in this specific area was apparent in his *Advice* where an "exquisite botanist" acts as apothecary in charge of the herbal garden attached to a teaching hospital. The practice of medicine is one special part of learning that Petty, trained in medicine at Leyden, sees as requiring immediate reorganization. Other particular parts of learning "where our own shoe pincheth us," according to Petty, were in the training of children, especially in vocational trades, and in the compilation of an encyclopedic history of all mechanical inventions and trades to date. This was needed before men could move forward. Too frequently, work was being repeated in ignorance of what others had already done. Hartlib's proposed Office of Public Address was to be a center for the collection, and Petty intended to support it, as well as his own utopian plans, from a spectacular tool for double writing, which he had recently invented and for which the patent had been awarded. He felt assured that from this source would come the necessary funds without which he distrusted his "flying thoughts" and those of others, which had collapsed from lack of the vast sums needed to carry out the designs. His *Advice* he thought feasible and practical; he had no "leisure to frame Utopias" and repudiated the fictional approach. The tract is certainly utopian in the new fashion, but a partial scheme, half-sister to the full-scale presentations in *Christianopolis* and *New Atlantis*. Petty explained that he was limiting it, since the definitions of learning and its advancement had already been "so accurately done by the great Lord Verulam."

The teaching hospital that Petty dreamed of in his *Advice* when he was only 25 years old was part of a larger medical center with administrative organization carefully delineated. Precise details were required for the patient's case history, and the staff was strictly trained to keep such records indicating which treatments were effective, and, if so, in what way—all most unusual in the 17th century when, for example, pigeons were still used to draw the vapors of illness from John Donne. Following his statistical bent Petty, years later, now Sir William Petty (he was knighted at the time of the Restoration), made his major bequest. He was the first to recommend the use of quantitative, empirical methods in the study of social and political phenomena and introduced the term "political arithmetick,"[26] which was the forerunner of 18th-century "statistics."

> Instead of using only comparative and superlative Words, and intellectual Arguments, I have taken the Course (as a Specimen of the Political Arithmetick I have long aimed at) to express myself in Terms of Number, Weight, or Measure; to use only Arguments of Sense, and to consider only such Causes as have visible Foundations in Nature; leaving those that depend upon mutable Minds, Opinions, Appetites and Passions of particular Men, to the Consideration of others.[27]

His *Political Arithmetick* continued, explaining lucidly the theory of applying exact measures to the study of society and advocating a proper census of inhabitants with precise records of age, sex, marital status, religion, and trade.

One suspects it was the figures in the last item—trade—that interested Petty most, and a check of his papers presented to the Royal Society, in which he was a very active contributing member, bears out the supposition: his favorite project urged upon the members was a history of trades, similar to that proposed in the early utopian *Advice*, and he gave numerous reports to the Fellows on developments in the clothing, dyeing, and shipping industries.

William Davenant and Abraham Cowley

Petty's *Advice* in 1648, which recognized the great Verulam's work as the basis for his own additions, just preceded the sharp rise in Bacon's prestige. Now at mid-century, with the imaginative colleges of Sir William Davenant (1651) and Abraham Cowley (1656), the College of the Six Days Works of *New Atlantis* reappears in close replica. These are two small scientific colleges, societies of scholars engaged in research and teaching and, oddly enough, placed in the midst of epic poems. Colleges of this type had certainly not been a traditional feature of classical epic or Renaissance romance. A comparison of these scientific centers with the Platonic, allegorical House of Alma in Spenser's romantic epic, *The Faerie Queene*, shows immediately the impact of science: the ethical emphasis of the House of Alma has shifted to laboratories in Davenant's House of Astragon, even as the magician's role in earlier romances changed to that of the natural philosopher in these 17th-century epics. It is positive evidence of the spread of the new doctrine when it invades the unlikely realm of poetry;[28] here it goes far beyond the draftsmen's offices of a Hartlib and Petty.

Furthermore, neither Cowley nor Davenant was notable in scientific fields. Although Cowley, a member of the Royal Society, knew and appreciated Harvey and Hobbes, Gassendi and Descartes,[29] he was in effect a popular bard of the movement, not an experimenter. Cowley's own contributions to scientific knowledge were limited to his six poetic books on plants, herbs, and trees, which his biographer Thomas Sprat saw as an outgrowth of his medical training at Oxford. His study of medicine was thought by some to be a deceptive maneuver covering his other activities as a representative of the King's interests in England during the Cromwellian period, but whether this was true or not, Cowley's generally intelligent understanding and interest in scientific development was certainly both sincere and extensive. In one essay he called for colleges of agriculture to be attached to Oxford and Cambridge, recommending some "industrious and publick-spirited" man like Mr. Hartlib, "if the gentleman be yet alive," to fill one of the chairs.[30] Yet there are no papers on specific research presented by Cowley at meetings of the Royal Society as there were in the case of Petty.

Davenant was wholly occupied with theatrical productions, operas with lavish settings and lady-actors, and royalist activities generally unrelated to the group of virtuosi. Still he saw fit to establish Astragon's scientific House in the world of his heroic poem, *Gondibert.* The hero wounded on the battlefield is brought to the wise and wealthy Astragon for cure. Astragon's skills obviously extend beyond medicine:

> This fam'd philosopher is Nature's spie,
> And hireless gives th' intelligence to art.[31]

To carry on his research, Astragon had a well-organized staff and excellent equipment for the laboratories. His scientific House is a vignette of the larger and more imposing Salomon's

House, which had preceded it by some 25 years and, like its model, Davenant's visionary college was discovering the hidden parts of nature by applying strictly inductive procedures.

In Cowley's small college, too, the new thought is avidly pursued, but with more emphasis on teaching. This time, in the epic *Davideis*, which Cowley probably wrote in his student days at Cambridge, there is a strange little utopian center of learning—the Prophets College at Rama—in the Biblical setting of ancient Judea. The heroic David, when fleeing from Saul's anger, finds refuge here in the peaceful simplicity of a religious and contemplative atmosphere. And though the prophets sing of God's creation of the earth and frequently meditate in their small tidy cells, an outstanding amount of their college's curriculum is devoted to the new philosophy.

In the college are three Old Testament figures, Mahol, Nathan, and Gad,[32] who teach, respectively, natural philosophy, astronomy, and mathematics. When the great Nathan explains the movement of the heavens, Cowley interpolates that though he is a prophet,

> his *Lectures* shew'ed
> How little of that *Art* to *them* he ow'ed.[33]

The author wants to make sure that no one thinks his prophets are "foretellers of future things." Rather, they are serious proponents of the new knowledge.

Because of the incongruous setting for the scientific attitude, the Prophets College attracts interest, but Cowley's later *Proposition for the Advancement of Experimental Philosophy* (1661) merits closer attention as a more complete utopian plan for scientific research. He installs laboratory equipment, assigns teaching and experimental responsibilities precisely, and envisages a general community of professors living together and working industriously to unearth "new mines" of knowledge. This carefully designed and detailed project is the work of a realistic author who has building costs in mind. Cowley asserts:

> We do not design this after the Model of *Solomon's* House in my Lord *Bacon* (which is a Project of Experiments that can

never be Experimented) but propose it within such bounds of Expence as have often been exceeded by the Buildings of private Citizens.[34]

Regardless of this criticism of Bacon's plan, Cowley proceeded to outline, in effect, a Salomon's House in which several features were identical with the original model, as will be shown. Furthermore, Cowley's goal for the scientific foundation was the same as Bacon's. Basically, the difference between the works lay in Cowley's more practical description of duties, expenses, and living arrangements for the Fellows.

So also the country gentleman to whom Cowley appropriately dedicated his essay on gardens, John Evelyn, abandoned Bacon's plan and then proceeded to suggest a scientific college noticeably resembling the Lord Chancellor's in several aspects. In a letter to Robert Boyle from Sayes Court, September 3, 1659, Evelyn outlined what he considered his more practical scheme, since "we are not to hope for a mathematical college, much less a *Solomon's* House," in such perverse times. Evelyn's utopian college need not be carefully analyzed in this study, because it was primarily a private, restricted club of six members only, who were to pursue their individual experimental projects in a congenial and most comfortable atmosphere; it was not a vision of organized research on a large scale for quick benefits to society. The significance lies in one more, and influential, adherent who believed and accepted the promise of science. Evelyn too was active in the Royal Society, serving as secretary in 1672, and he was twice begged to become president of the group. His published papers indicate the breadth of interest of a well-educated, traveled, and wealthy man: *Fumifugium; or the inconveniences of the aer and smoak of London dissipated, together with some remedies* . . . (a tract on the problem of air pollution familiar in our 20th century), a history of chalcography (engraving) to which was appended a new method, many works on gardening such as *A Discourse of Earth relating to the Culture,* which he presented to the Royal Society in April and May 1675, *A Letter to Lord Brouncker* [in the Society] *on a new Machine for Ploughing*, and *Numismata: A Discourse of Metals.*

So reads the assorted and fairly typical list of the publications of one man who, like many others, was operating in several directions at once and offering new, often crazy, but original ideas in the gestation period of science. There was no question of man's control over the new thought. The group following Bacon and his contemporary utopists was launching forth, freely applying the new knowledge, and watching with sharp eyes for solutions to everyday problems. They were not reformers in the sense that Campanella, Andreae, and Bacon actually tried to be; they were mainly concerned now with advancement of the work—the project—itself. No matter what field, it was still untilled and they could imaginatively project growth. Nor were they semi-philosophers attempting world reformation like Andreae and Campanella. Their support of the upsetting new doctrine, however, was psychologically and effectively telling.

The first historian of the Royal Society, Thomas Sprat, said that Cowley's *Proposition* greatly hastened the contrivance of the Society's platform. Cowley wrote the complimentary ode *To the Royal Society* introducing Sprat's *History* and, in this, he showed no reluctance in giving full credit to the Lord Chancellor for pointing out the way of the future:

> Bacon, like Moses, led us forth at last,
> The barren Wilderness he past,
> Did on the very Border stand
> Of the blest promis'd Land,
> And from the Mountains Top of his Exalted Wit,
> Saw it himself, and shew'd us it.

Men before had been like the wandering Hebrews in the desert; it was Bacon who had broken the "Scar-crow Deitie" of ancient authority and wisely advised "the Mechanic way." As a result, Cowley and his contemporaries in the Society are blessed:

> New Scenes of Heaven already we espy,
> And Crowds of golden Worlds on high.

They have, as he goes on in the ode to say, the telescope, so that "Nature's great Workes no distance can obscure," and the microscope, equally powerful:

> No smalness her near Objects can secure
> Y'have taught the curious Sight to press
> Into the privatest recess
> Of her imperceptible Littleness.

So the world will be contained in a grain of sand. Cowley saw the new discoveries in Bacon's exciting light as did Davenant, Evelyn, Petty, and Hartlib before them. The *New Atlantis* had been a "prophetick Scheam of the Royal Society."[35] There was no doubt in the minds of these early founders. Sprat wrote that there was

> one great Man, who had the true Imagination of the whole extent of this Enterprize, as it is now set on foot; and that is, the *Lord Bacon.* In whose Books there are every where scattered the best arguments, that can be produc'd for the defence of Experimental Philosophy; and the best directions, that are needful to promote it.[36]

In fact, Sprat continued that if he could have had his wish, "there should have been no other Preface to the *History of the Royal Society*, but some of his Writings." So Bacon took his place beside King Charles II on the frontispiece of the first history of the Society.[37]

R. H. and Joseph Glanvill

For the last two major utopias of the scientific type published before Newton's definitive work in 1687, Bacon's parentage was unquestionable. The reader will remember that he had left the *New Atlantis* unfinished, abruptly ending the work, which was not unusual in the utopian tradition: Plato had intended a "still larger design which was to have included an ideal history of Athens, as well as a political and physical philosophy," according to Benjamin Jowett (Introd., p. 18). Sir Thomas More had implied a further discussion of Utopia in the future, and

Samuel Hartlib arranged for another meeting between his two characters to continue their talk of Macaria. Utopias are never finished; man can always dream on. Within this tradition Bacon's publisher, William Rawley, said the Lord Chancellor had planned to draft model laws for Bensalem, but felt his time might be better spent collecting data for the great "Natural History." Thus the *New Atlantis* was left a fragment, but not for long.

In London in 1660 there appeared anonymously[38] the *New Atlantis. Begun by the Lord Verulam, Viscount St. Albans: and Continued by R. H. Esquire. Wherein is set forth A Platform of Monarchical Government. With a Pleasant intermixture of divers rare Inventions, and Wholsom Customs, fit to be introduced into all Kingdoms, States, and Common-Wealths.* Sixteen years later Joseph Glanvill presented another "Continuation."[39] Each begins with a brief summary of Bacon's story and purports to complete it. As R. H. explains, he wishes to enlarge and "add one cubit more to that rare modell of perfection" which the "Princly architect left unfinished." It is honor enough for him to "carry a torch behind so great a light" as Bacon's work, which is the solid foundation to which he would add the "higher floors."

R. H. creates a most imaginative and formidable superstructure for the *New Atlantis*. Continuing the first-person narrative of his predecessor, he adds the beautiful city of Bellatore with Christ's Church decorated by paintings of Titian and Michelangelo, three large universities located in major cities, an educational program for children, a monarchical platform of government, and many interesting customs that Bacon had neglected to mention. The result of his additions is an improved Bensalem in which all citizens are vocationally trained and skillful in using the new knowledge for practical purposes. Since Bacon had elaborately set forth the inductive method of research, R. H. chose to emphasize the final product—technical inventions and science applied for man's benefit. He was a true heir of Bacon's utilitarian philosophy, displaying it in a most vivid way.

In contrast to R. H.'s stress on applied science, Joseph Glanvill describes the ideal intellectual atmosphere for the popular community of Bensalem, and remains consistently faithful to the subject announced in the title: "Anti-fanatical Religion, and Free Philosophy. In a Continuation of the *New Atlantis*." He prepares the reader by stating that his essay is "a mixture of an Idea, and a disguised History," for which he has "borrowed the countenance, and colour" of the Lord Chancellor's story. After the introduction, however, Glanvill is not concerned with the color of the story. There is neither action nor visual description; the essay merely records the conversation of the Governor and the author, who visited the community. It is evident that Glanvill used the utopian form only for convenience in expressing his philosophy.

When he reviews Bacon's "history" of the *New Atlantis*, Glanvill calls Salomon's House a "Royal Society erected for Enquiries into the Works of God." The empirical method employed by the staff workers of Salomon's House and by the Fellows of the Royal Society is the "right method," according to Glanvill, a prominent defender of the Society's work when it was attacked by Henry Stubbe and others. But Glanvill was also a conspicuous spokesman for liberty of inquiry and the open-minded consideration of all philosophic theories. Thus the Bensalemites, created by Glanvill in the year 1676, were thoroughly engaged in the rational examination of various systems of philosophy current in the intellectual world of their time. Their attitudes constitute an excellent summary of the developmental stage of modern philosophy just a decade before the first book of Isaac Newton's *Philosophiae Naturalis Principia Mathematica* was exhibited at the Royal Society.

The Group and the Literary Genre

Glanvill, trained at Oxford, was a minister in the Church of England and became Chaplain in Ordinary to His Majesty Charles II; Cowley, Davenant, and Evelyn were all royalists in the Civil War period and generally best typed as gentlemen of letters without other specific professions; Hartlib had been an intelligencer in the real sense of broadcasting the new thought;

and Petty's reputation was made as a demographer for future generations. The avant-garde of scientific utopists were all well-known figures in their own day, but few, with the exception of Bacon, have come down to us as great men. They were the second line, neither the original thinkers nor lawgivers in science, but without their dissemination and projection of the doctrine, the acceptance of scientific discoveries would undoubtedly have been slower.

In the social hierarchy they were not aristocrats, though Petty and Davenant were knighted and Bacon received the titles of Baron Verulam and Viscount St. Albans. Generally these utopists represented the active and expanding middle classes of society. In the early decades of the century, science was not considered a suitable pursuit for the gentleman. Sir Kenelm Digby, an eager amateur but poor contributor, commented correctly in the 1630's that a scientific interest was new to a "man of quality." William Robinson, writing to Oughtred in 1633, complained that science was slighted and held in little esteem. And Cowley, years later, hailed Digby as a protector of the "arts and sciences in an age when they are being contemned." Although Cowley and other defenders continued to lament the lack of widespread support, interest had consistently increased, reaching a peak from 1646 to 1650 and a high point in actual discoveries and inventions in the 1660's,[40] the same period in which Bacon was lauded as the guide to the Promised Land. By this time King Charles II gave his patronage, even experimented a bit himself, and chartered the Royal Society in 1662. "Gentlemen of culture" joined the pursuit as minor virtuosi with the blessing of the monarch (but no financial support). Soon it became a parlor game to experiment, and ladies cherished little microscopes as amusing toys.

The position of science and its concomitant technology was assured by the last quarter of the 17th century. Authors like Andreae, Campanella, and Bacon had written with effective pens in hand when promulgating the new ideas. The empirical philosophy exemplified by Salomon's House was patently the model for utopian colleges from Davenant's copy on through Glanvill's "Continuation." As a group, regardless of specific influence, these men were highly sensitive to the climate of opinion (Glanvill's term), quickly stimulated by latest developments, and they rapidly generated proposals on how to make the knowledge useful.

The genre of utopias, like all such literary patterns, had been changed also in accordance with the goals in mind. Of course future utopias saw again purely imaginative sketches of the better life, but these authors of the 17th century often used the structure while denying the intent: they would be more realistic and practical, not utopian dreamers, because their schemes were to be operative tomorrow. The illusion of authenticity that had characterized many older utopias was often abandoned or forgotten in utopian proposals later in the 17th century. Sir Thomas More had employed imaginary "facts" most amusingly when he explained that his (fictitious) informant, Ralph Hythloday, whose name means "huckster of idle talk," had been left behind at his own request when sailing on the last voyage of Amerigo Vespucci. The credibility of his story was increased, since a report of Vespucci's travels had actually been printed less than ten years before More's *Utopia* in 1516. Following this general pattern, Andreae had created a shipwreck that placed him in Christianopolis, where he could have the "realistic" experience of touring laboratories and the city. Campanella had reported the dialogue between a Genoese sea captain and a Knight Hospitaller. The captain had simply been "compelled to go ashore" and found himself in the land of Taprobane, "immediately under the equator." Bacon's seamen discovered the island of Bensalem by chance when driven off their course at sea. Hartlib had retained the storyteller's touch when he arranged for his scholar to meet a traveler and "take a turn or two" in the Moorfields as they discussed the sensible solutions of problems in Macaria.

But Hartlib presented his utopia to Parliament for support and others, like Petty and Cowley, were concerned with finances for the work to be done. The utopian setting of Hartlib's *Macaria* was only to make the outline palatable and entertaining —to delight while instructing. Examination of succeeding utopian plans reveals a pronounced tendency to abolish the fictional element altogether: Petty's *Advice* and Cowley's *Proposition* were written in straightforward essay form, and "utopia" applies only to the idealistic nature of the projection—it becomes the descriptive adjective "utopian." They are no longer interested in tricks of first-person narrative as used by Andreae, Bacon, and R. H. in his faithful completion of *New Atlantis.* Nor are they tempted to follow the dialogue deriving from Plato's *Republic* and employed by Campanella and Hartlib. Their rejection of fiction was related to the growing distrust of imagination and make-believe worlds. Cowley's ode *To the Royal Society*, in which the vistas of scientific achievement are most happily viewed, renounces the sterile "Desserts of Poetry," the "painted Scenes, and Pageants of the Brain," which have misled men while the riches lie hoarded "In Natures endless Treasurie." Again, in his ode *To Mr. Hobs*, who is the "great *Columbus* of the *Golden Lands* of *New Philosophy*, it is his solid *Reason* like the *shield* from Heaven" given to the Trojan hero that wins Cowley's acclaim. The age of proof and reason was beginning.

As part of the same attitude, no ornament was allowed in style; flowing rhetoric and eloquence of the high style were thoroughly out. The scientific utopists, many of them members of the Royal Society, agreed with the Society's effort to speak clearly and directly, to fashion the English language for its greatest effectiveness upon the growing number of readers. So the plain style is typical of these authors, matter-of-fact and familiar. The colloquial and picturesque example replaces the learned reference. They are among the fathers of the great English middle style that proved serviceable in all types of writing.

Thus when the scientific attitude entered the utopias following those of the early triumvirate, it had far-reaching effects even upon the style and literary format of utopias. But more importantly, the projections of the future and the possibilities for science, especially in its applications, which were created by these imaginative men, helped to spread the doctrine and determine the utilitarian philosophy that accompanied the discovery of natural laws.

VI

UTOPIAN ANSWERS TO AUTHORITY

THE ENTRANCE OF MODERN SCIENCE into utopia was no smoother than its emergence in the real world. Few in the audience joined the utopists in their enthusiastic response; instead, it was attended by some severe critics who reacted strongly against the claims and even the very existence of science. This the new utopists recognized and promptly accepted the challenge, knowing full well that their proposals could not be convincing until they had destroyed the now old-fashioned molds that still determined beliefs and values for the majority. The key arguments of their vociferous and entrenched opponents, who subdued man's curiosity about the new world of nature by asserting the authority of tradition, had to be refuted before science could rapidly progress either in the real or ideal society. For the impatient reformers there were three major authoritative obstacles: (1) established educational institutions, (2) belief in the superiority of the ancients, and (3) church doctrine.

Educational Institutions

In opposition to the first two closely associated obstacles, the utopists as a group generally advanced the same arguments. They engaged in the battle with current educational practices, using every weapon at their command, and set up programs—quite novel for the 17th century—to educate youth. In fact, interest in this area of reform became the main concern in some utopian tracts like Cowley's *Proposition*, which was completely devoted to an ideal scheme for the promotion of learning, after which, it was understood, all society would be renovated. Other utopists had more difficulty separating the educational approach from so-called scientific methods that were not yet clearly defined. The advancement of natural philosophy rightly embraced both; so the curriculum and teaching plans often were almost indistinguishable from laboratory procedures, as in Andreae's easy transfer between the classroom and the experimental research program. Similarly, the drive for more knowledge of the physical world, which had practical implications for the utopists, appeared in their emphasis on vocational training and active employment. Learning was to result from experience, which was Hartlib's and Petty's thesis, as well as the formal classroom. And while Cowley, like Milton in his tract on education, would introduce the student to natural sciences in the classroom through the best-known Latin authors, it was for the purpose of differentiating truth from falsehood in the ancients' writings and then proceeding to "improve all Arts" and to "discover others which we yet have not."

Some were not so tolerant of the past or failed to recognize a heritage: only the practical, immediate field of work would yield the new information. Such was the attitude of Gerrard Winstanley, who put forth his ideas in *The Law of Freedom in a Platform or True Magistracy Restored* (1652).[1] Winstanley was an independent spirit who stood apart from the other utopists thus far considered. He was less the philosophical student of affairs, more the fiery man of explosive action. In the years just preceding the publication of his utopia, he had been a leader of the Digger Movement, instigated by a group of radicals espousing communistic, economic reforms, and he had been involved in serious altercations with rich landholders and government officials as well as with the clergy, which no doubt increased the extremism in his projection of the ideal state. Although his tract was sharply critical on contemporary economic and social issues, it still provided a well-balanced statement of the general attitude toward science in England at this time. While bravely championing the underdog in his generation, Winstanley had absorbed the new thought.

The *Platform* was presented to Oliver Cromwell and obviously intended to serve him as a guide for the new order in process of formation. For this reason, all fictional setting was ignored and the ideal society described realistically in the strong, vigorous prose of pamphlet style. Winstanley's plan was unquestionably novel in its day and was accompanied by vitriolic comment on the current educational program. All men in his perfect commonwealth, which took for its ideal Israel because it was considered the original example of righteousness and peace in the world even as Harrington had studied it as one of his models,[2] were brought up "in every Trade, Art, and Science, whereby they may finde out the Secrets of Creation." In due course, "there will not be any Secret in Nature, which now lies hid . . . but by some or other will be brought to light." Thus he affirmed his faith in the effort to understand the universe and its laws.

As the basis for the curriculum he suggested "five Fountains from whence all Arts and Sciences have their influences:" (1) husbandry and medicine, or "what is good for all bodies, both men and beasts;" (2) "Mineral employment, and that is to search into the Earth to finde out Mynes" of all sorts; (3) and (4) the "right ordering" of cattle and timber resources; and (5) the study of "Astrology, Astronomy, and Navigation, and the motions of the Winds, and the causes of several Appearances of the Face of Heaven, either in Storms or in Fareness." Only in the last fountain of knowledge is reason mentioned as necessary to the process of discovering the unknown. The other four areas of study rest completely on observation and experience. Winstanley will not have a "studying imagination" or dependence on words and systems of logic, for these are the "devil . . . who puts out the eyes of mans Knowledg, and tells him, he must believe what others have writ or spoke, and must not trust to his own experience." Winstanley, like Petty, Hartlib, and Campanella—indeed the majority of new utopists—placed his trust in experience and the senses, the empirical approach to knowledge, not in what others have recorded in books. Bacon's was a notable and constant attack on the universities as barren places of knowledge. He denounced the accepted ends of education, the methods and exercises employed, and consistently asked for a new institution in which natural science and mechanical arts could be advanced.

The main charges against contemporary education brought by the optimistic reformers were aimed at words, rhetoric, and disputation; they were dead set against the prevailing spider web of scholastic reasoning. Their emphasis was on the concrete object and experimentation with due regard for sensory powers. Consider the revolutionary distance between this approach and the typical university curriculum of the period with its public orations presented in Quintilian style on some subject such as Milton wrote on during his career at Cambridge: namely, whether "Learning Makes Men Happier Than Ignorance," "On the Music of the Spheres," or "Whether Day or Night is the More Excellent." It becomes ridiculously clear at once that such topics could scarcely be of interest to those who would make a new world with science; in fact, such subjects were a futile waste of time, training young minds incorrectly and diverting them from the more important new methods and work to be done.

Criticism of universities bcause of their adherence to the scholasticism of the late Middle Ages had begun with the humanists and reached a high pitch of fervor in 17th-century England. Vittorino da Feltre and others over a hundred years earlier had established schools with a new atmosphere, not necessarily scientifically oriented, but at least they had dispensed with the rigid curriculum of their predecessors and created a happy schoolroom in which children could laugh and play while learning. This feature was characteristic of the new utopia's school and still it was considered experimental and "progressive" because Vittorino's model had neither been widely accepted nor copied in the intervening years. In utopia we entered a classroom that was open, sunny, and beautiful with not a trace of the gloomy conditions of learning in the real world. Teachers adapted work to the individual's talents and abilities, and treated pupils kindly and courteously. There was no desk for a stern schoolmaster who might administer the ruler or cuff the recalcitrant learner.

Instruction was in the mother tongue and modern languages

had a prominence equaling, even surpassing, Greek and Latin. Languages were best taught by use and experience, with reading materials planned for the child, and an incredible number of languages were to be quickly mastered. This was to open all doors to knowledge, current and past, though the portal to natural philosophy was first in importance. Andreae stressed the study of many languages for Christianopolitans just as he had personally perfected them in his own life. Campanella, however, took a somewhat more practical view: his most educated man, the high ruler of Solarians, did not need to spend much time studying languages, because many interpreters were available and "grammarians of state" were on hand for public pronouncements. Despite Campanella's specialized services, languages were generally considered vital in the new curriculum, and modern tongues were as respectable as dignified Latin, the universal language of scholars.

In the instructional process, the visual approach was singled out as most effective for teaching, not visual obviously in reading the written word and copying it, but visual in the sense of contact with the object itself being studied. Solarian children were led around the walls of the *City of the Sun* while a highly revered teacher explained the purposes and habits of each object visually before them. Comenius produced the first pictorial instruction book for children on which his reputation with educators throughout the years has largely rested. Holding the same opinion, Andreae spoke definitely for the efficacy of visual learning:

> Truly, is not recognition of things of the earth much easier if a competent demonstrator and illustrative material are at hand and if there is some guide to the memory? *For instruction enters altogether more easily through the eyes than through the ears.*[3]

In this passage he was describing the method used in the natural science laboratory, which was no different from the method of classroom study on the same subject. And Glanvill, too, had both the educational and the laboratory method in mind when he accused the scholastics of having left the world "*intellectually invisible*" by their dry generalities and wordy discussions of principles and influences. Now his generation would "*visualize* the hidden processes of Nature," they would draw "intellectual diagrams" to make visible what their forebears had left buried in the unknown or the occult.[4] Visual emphasis was a strong part of the new utopists' answers to scholastic influence in educational institutions.

Similarly, field trips, actual practice, and the use of the hands in crafts, industries, and agriculture were designed for utopian children to replace the medieval system that prevailed in non-utopian societies. In utopia they learned to name and classify thousands of herbs simply by playing joyously among them; John Dewey's thesis of "learning by doing" was typical of the imaginary practices of the 17th century. The classroom was not a place where the student passively received the wisdom of the past. Rote memory was abolished as a means of instruction, for children were no longer treated as sausages to be stuffed with knowledge by the teacher and then tied up with the ribbons of the diploma. Memory was not to be a "Sepulchre, furnished with a load of broken and discarnate bones" from the past.

Sir William Petty carefully planned the new program in his *Advice:* he withheld abstractions from beginning students and adapted the teaching process to the child's natural interest in objects and his powers of imitation.[5] The system was geared to the levels of maturity, because the student was considered a developing organism who learned by copying his elders in the elementary phase. In this manner, all children, regardless of social rank, were taught "some genteel manufacture in their minority," such as the making of astronomical, musical or mathematical instruments, modeling of houses and ships, or stone-polishing. The value of this learning to adult life was clear, and whether they became laborers or rich patrons, they were to support the great design of advancing knowledge. Petty thought of a human being as a tool, an instrument to be whetted and sharpened by education for carrying on the new work of science. His thinking thus, however, had no negative connotation as we might now read into it because of the suspicion that we are perhaps tools of the mechanized society. Petty simply saw the classroom as a laboratory, and the end product as a scientist or technician equipped to further learning.

While utopists advocated education for the scientist, Oxford and Cambridge were training divines and statesmen as they had for many years. This meant that the man of science was generally self-taught. Chemistry, mathematics in its newer developments like Napier's logarithms, and such facilities as laboratories for instruction were lacking within the universities' quadrangles. Throughout the century educational institutions were generally aligned with the conservative, majority opinion, which has most often been true of the protectors of the heritage in human knowledge. So curricula in the English universities remained in old patterns that were adjusted slowly, too slowly for men inflamed by the possibilities of the new philosophy.

John Hall, who was neither a scientist nor a utopist but a courageous commentator, stated the complaint in 1649 in *An Humble Motion to the Parliament of England Concerning the Advancement of Learning:*

> We have hardly professours for the three principall faculties and these but lazily read,—carelessly followed. Where have we anything to do with chemistry which hath snatcht the keyes of nature from the other sects of philosophy by her multiplied experiences? Where have we constant reading upon either quick or dead anatomies, or occular demonstrations of herbes? Where any manuall demonstrations or mathematicall theorems or instruments? Where a promotion of their experiences which, if right carried on, would multiply even to astonishment?[6]

The line of questioning is typical of Comenius, Hartlib, and John Dury in the 1640's and 1650's, when they fostered many plans to correct the educational situation.

In those relatively rare instances in which a subject of interest to the scientist was offered by the university as, for example, mathematics, which was always present in the university program, the experimentalist still objected to the method of presentation, the antiquated approach, or the lack of the latest materials and information. John Wallis, the famous mathematician and one of the Royal Society's founders who had met with William Petty in the early group at Oxford, still lamented at the end of the century that there had been no one to direct his study of mathematics, at least not in the way he thought it should have been taught.

> I had none to direct me, what books to read, or what to seek, or in what Method to proceed. For Mathematicks, (at that time, with us) were scarce looked upon as *Academical Studies*, but rather *Mechanical;* as the business of *Traders, Merchants* . . . and perhaps some *Almanack-makers in London.* And amongst more than Two Hundred Students (at that time) in our College, I do not know of any Two (perhaps nor any) who had more of Mathematicks than I (if so much) which was then but little . . . for the Study of *Mathematicks* was at that time more cultivated in London than in the Universities.[7]

Individuals furthered their independent study with the assistance of tutors and by associating with men of like interests in informal groups, which developed into academies similar to those in Italy and France.

It is all too common, however, for critics to describe the universities' situation in this period with a general statement about the "lag" and tendency to preserve outmoded patterns. Steps to modernize the curricula were taken, though often they came from outside influences as, for example, the establishment of various "chairs" in the newer subjects. At Oxford, the Savilian professors in geometry and astronomy (1619) were allowed an unusual amount of freedom. Under the terms of the grant, the professor of astronomy was permitted to discuss the Copernican hypothesis after he had given due attention to the Ptolemaic system, and he was expected to conduct experiments. Yet it was many years until the appointed professor took advantage of the privilege to advance new theories. One reason no doubt was the lack of textbooks and popular writing explaining the Copernican thesis simply for the beginning student or intelligent layman. Only a passing sentence or two was given it until Blaeu's *Insti-*

tutio Astronomica appeared in its Latin translation by Hortensius in 1668, and even then Blaeu started with the Ptolemaic system first, so that the beginner would have less difficulty understanding the newer theory.[8] So the problem was compounded by lack of instructional materials.

In 1621, Sir Henry Savile's son-in-law established the Sedleian chair in natural philosophy but, aside from these innovations, there was little activity toward reform in the curriculum until the Commonwealth, when Puritans entered the faculties in greater number and a more liberal attitude toward the new science and practical education appeared. Outstanding men like John Wilkins, master of Wadham College when the young scientific group centered there, Seth Ward, and Wallis were all teaching at Oxford in this period, and changes did occur. Ward was first in actually presenting Copernicus' theory in 1649 and Wallis, who held the Savilian chair in geometry for many years, 1649–1703, received credit for introducing Descartes' analytic geometry.[9]

Although chairs in chemistry and botany were added in 1659 and 1669, respectively, Oxford did not retain its lead over Cambridge University, which captured supremacy in the physical sciences during the latter part of the century. Joseph Glanvill regretted that he had not studied at Cambridge, where the new philosophy was held in more esteem. Cambridge had housed the interesting Neoplatonist group including Benjamin Whichcote, Ralph Cudworth, and Henry More, whom Glanvill greatly admired. In 1663 the Lucasian chair there was filled by Isaac Barrow and next by Isaac Newton. Even so, during the last two decades, chemistry was taught by the Frenchman Vignani outside the Cambridge curriculum.[10] The new learning certainly had had its effects, but on the whole it did not come quickly to the universities, which were bound by custom and statutes[11] as well as traditional ideas.

Regardless of curricular adjustments, the new utopists considered the entire educational pattern on the wrong foundation. Abraham Cowley, in the preface to his *Proposition*, summarized the attitude of all utopians toward established educational institutions:

> The Inquisition into the Nature of God's Creatures, and the Application of them to Humane Uses (especially the latter) seem to be very slenderly provided for, or rather almost totally neglected, except only some small assistances to Physick, and the Mathematicks. And therefore the Founders of our Colledges have taken ample care to supply the Students with multitude of Books, and to appoint Tutors and frequent Exercises, the one to interpret, and the other to confirm their Reading . . . that the Beams which they receive by Lecture may be doubled by Reflections of their own Wit: But towards the Observation and Application, as I said, of the Lectures themselves, have allowed no Instruments, Materials, or Conveniences.[12]

The utopists were interested in instruments and materials, not book-learning and involved debates on fine points of logic as taught in the university. Subject matter in utopia was no longer the trivium and quadrivium of the traditional liberal arts curriculum; instruction was not intended to train the polished courtier who must charm with witty conversation on love; and courses were not designed after the pattern of Cambridge and Oxford for the education of statesmen and clergy. Nothing new was to be discovered from these conventional paths. Instead, utopists installed laboratories in the classroom and planned an educational program based on observation and practical, firsthand experience, particularly in the various branches of natural philosophy. Both the curriculum and the method of teaching were radically changed in the transfer from the real to the imaginary school.

Superiority of the Ancients

The second obstacle toward which the utopists passionately directed charges was implicit in the first: namely, the high regard for the classics in the official curriculum. Humanistic scholarship had greatly advanced the ancients' position, and classical authors reigned in university studies and cultural life as well. Ancient writers were the authorities in all matters of learning, and consequently a source of discouragement to the moderns' efforts. What more could be discovered if the last irrefutable word had been spoken?

Again, Cowley was a succinct representative of the modern utopists when he objected to the

> pernicious opinion which had long possest the World, that all things to be searcht in Nature, had been already found and discovered by the Ancients, and that it were a folly to travel about for that which others had before brought home to us. . . . Not that I would disparage the admirable Wit, and worthy labours of many of the Ancients, much less of Aristotle, the most eminent among them; but it were madness to imagine that the Cisterns of men should afford us as much, and as wholesome Waters, as the Fountains of Nature. . . . And no man can hope to make himself as rich by stealing out of others Truncks, as he might by opening and digging of new Mines.[13]

As additional evidence in favor of the moderns, Cowley points out that the ancients never even dreamed of America, and he adds reassuringly that "there is yet many a Terra Incognita" to discover. The new method of education and research was to defeat the ancients by producing results of benefit to all, and by revealing new truths.[14]

This question involved the final confutation of the old theory of decay in the universe, poetically expressed by Hesiod in his five stages of civilization declining from the Golden Age, and supported by the Christian belief in the fall of man from his first estate of perfection in the Garden. Such depressing doctrine had to be reversed if one were to have faith in the future. Here Bacon's optimistic trumpet was instrumental, pronouncing the epitaph for the old view and turning men's sights forward, establishing belief in the idea of progress.[15] Bacon maneuvered by paradox, not condemning the ancients, but showing the accumulation of knowledge that placed the modern—dwarf though he may be by comparison—on the ancient giant's shoulder and enabled him to see farther ahead. All the scientific utopists, from Johann Andreae down through the century, were firmly on the side of the moderns in this battle. With Campanella, they agreed that their own age had in it "more history within a hundred years than all the world had in four thousand years before!" And the near future promised the greatest advances civilization had ever known.

In their rejection of ancient authorities, one figure loomed large as the main object for destruction. This was Aristotle, clothed by his many interpreters, and the central figure in the comprehensive system of philosophy inherited by the new utopists. Frequently they reacted with violence against Aristotelian thought and the schoolmen who had upheld it. Even Samuel Gott's inhabitants of *Nova Solyma*, tentative about accepting the new science, sharply opposed peripatetic philosophy. Throughout the many intervening years Aristotle's writings had received numerous interpretations and been used to support a wide variety of theories. Churchmen and philosophers had mixed his thought with their own, which often resulted in a hybrid of ideas untrue to Aristotle's statement. The schoolmen of the late Middle Ages had thoroughly adopted him as their major authority. Hence the 17th-century reader knew Aristotle generally from secondary sources, not infrequently twisted in meaning.

Intelligent citizens of the first new utopias were often so rabidly against Aristotelian influence that they were unwilling to consider his writings in the original and see how he had been distorted and misused. Andreae's compliment to Campanella was for his indefatigable fight against Aristotle, and certainly Bacon was clear when he rejected dependence on past methods and authority. He, however, compared his approach and belief with the ancients individually: he was kinder to Plato because of his induction and concern with forms than he was to Aristotle; he definitely favored Democritus for his atomic theory; and he praised Hippocrates for his effort to keep medical case histories and to evaluate the efficacy of various treatments for illness. Still, his general distrust of words and rationalism alone was most obvious, and it was reflected and repeated even more strongly by his close followers. Later in the century, utopists were more calm and considerate in dealing with this opponent: Abraham Cowley voiced respect for Aristotle as the most eminent of ancients and counseled the students in his utopian college to read especially the *History of Animals* and the writings on "Morals and Rhetorick." This was no threat to the new mines of knowledge to be opened. And Glanvill also provided a more reasonable intellectual atmosphere for the improved *New Atlantis*. Though he was adamantly opposed to scholasticism, vain disputes, and any abstract notions unassisted by the senses, he allowed a rational evaluation of Aristotle's concepts and even his method, which had been most distasteful to Bacon. By the time of these latter utopias, the study of syllogistic logic was recognized as valuable, at least in training young minds to think carefully. But, of course, one did not stop there. Although some ideal citizens were more vehement than others in their rejection, none permitted the shade of Aristotle to dominate thought in the perfect society.

The attack on Aristotle was not confined to educational matters. As the curricula based on training in disputation and logic ceased to satisfy the questioning new mind, the general prestige of the ancient, Aristotle, was challenged in other areas as well. The intricate, complex structure of the universe and its purposive meaning had been solidly set forth in the Middle Ages; it had remained effectively unchallenged for some 300 years. One idea interlocked directly with another to form the whole cohesive philosophy; so when one piece of the total structure was displaced, the remaining parts ceased to carry full meaning and their function was questionable. Thus, as Glanvill indicated in his description of opinions in Bensalem, when the Aristotelian physical universe was disproved, it was no longer of any value to the religious theory it had upheld.

Replies to Church Doctrine

In this way the attack on existing authorities was extended into the vital area of church doctrine, which was hotly debated in the

17th century and was closely related to many utopian proposals. The issue lay not in faith or belief in God, but more particularly in church doctrine or how men have interpreted God's purposes and intentions for the believer. No one of the new utopists doubted God's powers, but the extent of His determining hand had to be re-examined in view of natural laws that appeared also to be determining and almost mechanistic. The simplest and most frequent solution was to equate God's powers with natural laws; hence the conflict could be resolved and man, studying physical phenomena, was indirectly increasing his knowledge of God as the great architect, the deviser of the remarkable system. But the opposition would not rest content with this conclusion: church spokesmen objected to the study itself and desperately feared man's gaining a foothold of knowledge in the realm of God's designs. There was nothing new in this argument except the heat generated by the new scientific discoveries that were forcing the argument to a final show-down.

At heart, it was the same issue Enoch had raised when he inveighed against man's presumption; it was the same question Sir Thomas More faced when he justified man's curiosity in terms of the greater glory that would come to God from man's learning of His secrets; and it was the point on which we saw Samuel Gott part company with the more scientifically oriented utopists. While Gott's ideal people recognize that the "special advantage of natural science is to rise from nature to nature's God, tracing his footsteps everywhere therein," they also realize the danger of letting the search run away with one. It was becoming so all-absorbing that man might forget the honor was due to God and not to man himself, feeling his prowess in new discoveries. For this reason Gott had cautioned and controlled the limits for such research in his New Jerusalem. The insurgent doctrine of individual responsibility and excellence, reinforced by admiration for the Greeks, had gained many adherents and encouraged them to believe increasingly in their own powers.

Among the disciples urging man to employ all his abilities were the new utopists, who directly disagreed with Enoch's point of view and who questioned the validity of such positions as Dante's when he wrote that

> He is mad who hopes that reason in its sweep
> The infinite way can traverse back and forth.[16]

The infinite was changing in meaning: no longer an attribute of God alone, space as a physical entity was infinite, and time was joining infinity in the sense of belief in an unlimited future. The theory of progressive degeneration and decay of the world to the point of its final dissolution or conflagration had been repudiated by the new thought.

The task of each utopist, then, was to work his own way to a unified philosophy that combined the new ideas and concept of progress with a belief in God, His predestination, and His providence. Generally, they justified their faith in man and his determining abilities as evidence of his use of reason, God's greatest gift to him; providence and predestination were gradually redefined to support belief in progress as evolutionary improvement *ad infinitum*, evidence of God's beneficence; and God's providence was further revealed in His thoughtful provision for the universe and its intricate workings. So church doctrine, including God's role, merely required adjustment to accommodate the facts, and new utopists continued to live in a whole, comprehensive world of order.

But in this complicated issue lay another extremely basic and more difficult question. More's contemporary, Cornelius Agrippa, had stated it most clearly and it endured as a fundamental argument on the side of the church a hundred years later:

> For all the secrets of God and Nature, all manners of Customs and Laws, all understanding of things past, present, and to come, are fairly taught in . . . the holy Scripture. . . . Why seek ye knowledge of them, who having spent all their days in searching, have lost all their time & labour, being unable to attain to any thing of certain truth? Fools and wicked men . . . strive to learn from lying Philosophers, and Doctors of Errour, those things which ye ought to receive from Christ and the holy Ghost![17]

All things were taught in the Bible alone and only the "enlightened" or chosen few might hope to understand the parables and riddles sealing the knowledge safely away from the ignorant multitude. Agrippa raised a vital question: Wherein lay the source of truth? All, he said, was to be found in Scripture and there alone. What of the laboratory and empiricism? Did not the nature of the materials being examined differentiate among the methods to be employed? Abraham Cowley thought so and, as late as 1661, he felt it necessary to answer the challenge before outlining his ideal college for experimental philosophy. On this subject his preface starts:

> All Knowledge must either be of God, or of his Creatures, that is, of Nature; the first is called from the object, Divinity; the latter, Natural Philosophy.

With a stroke of the pen, Cowley separated theology and natural philosophy as two parts of learning and implied no conflict. They were simply discrete areas of knowledge and both equally part of God's scheme of things. For some utopists, however, the line of demarcation was not so clear and comfortable. Doubt remained about which was the highest truth.

It is a long-standing argument between those who find ultimate truth only in religion and those who claim truth only for observed and tested natural phenomena. The issue for some people has never been settled; instead, we travel two roads to truth as Douglas Bush describes it.[18] And occasionally there have been periods of sharp warfare over the question, set off by a particular discovery or an accumulation of new facts. This is what happened in the early 17th century, when many new facts were learned and proof was presented by Galileo supporting the Copernican hypothesis, which in turn was declared heretical by the Roman Church and branded as false in 1616—just six years after Galileo's evidence published in the *Starry Messenger.*[19]

The details of the religious argument at this time against the new learning are presented concisely in Campanella's *Defense of Galileo*, in which the author supports the revolutionary doctrine, or at least its right to exist. In the first spirited chapter he lists numerically the charges against Galileo:

1. He has completely overturned theological dogmas, and attempts to introduce new theories opposed to the science and metaphysics of Aristotle upon which Saint Thomas and all scholastics have based theological doctrine.

2. He promulgates opinions which contradict those of all the Fathers and scholastics. Indeed, he teaches that the Sun and sphere of the stars remain fixed, that the earth moves, and is placed beyond the center of the world. The Fathers, the scholastics, and our senses inform us differently.

3. Galileo openly contradicts Sacred Scripture. It is said in Psalm 92, "For he hath established the world which shall not be moved," and in Psalm 103, "Who has founded the earth upon its own bases; it shall not be moved for ever and ever." And Solomon states in Ecclesiastes 1, "But the earth standeth for ever."

. . .

8. Galileo also says that water exists on the Moon and the planets, which cannot be. These bodies are incorruptible, for do not all scholastics contend with Aristotle that they endure without change throughout all time? He described lands and mountains in the Moon and other celestial globes, and not only vilifies immeasurably the homes of the angels, but lessens our hope regarding Heaven.

. . .

11. Holy Scripture counsels us to "seek nothing higher, nor attempt to know more than it is necessary to know"; that we "leap not over the bounds which the fathers set"; and that "the diligent searcher of majesty is overcome by vain-glory." Galileo disregards this counsel, subjects the heavens to his invention, and constructs the whole fabric of the world according to his pleasure. Cato rightly taught us to "leave secret things to God, and to permit Heaven to inquire concerning them; for he who is mortal should concern himself with mortal things."[20]

If "new philosophy" is substituted for Galileo's name in order to broaden the implications, the first charge is certainly correct: orthodox religious dogma was overturned on many important and basic counts. The old Ptolemaic–Aristotelian cosmology had served as a prop sustaining the definite location and accepted characteristics of heaven. The outermost heavens had, for example, contained the palatial mansions intended for saints. Where were they to live when the old theory was removed and established theological doctrine collapsed? Churchmen—both Protestant and Catholic—countered with literal citations of the Holy Scripture bulwarked by the Church Fathers, Aristotle, Aquinas, and scholastics to prove the new astronomy wrong.

On the eighth count, Galileo had explicitly claimed that he would prove not only the earth to have motion and surpass the moon in brightness, but that the earth "*is not* the place where the dull refuse of the universe has settled down," thus directly challenging the Aristotelian notion that only in sublunary areas was there change and mutation, that above the moon all was stable and fixed in perfection which, of course, protected heaven's environment. And on the last eleventh point, we hear again the voice of Enoch: man was not to leap over the established bounds. This argument was zealously revived to discredit experimenters on charges of "vain-glory" and presumptuous attempts to enter Deity's sacred realm.

Campanella's defense was an extremely able argument, well controlled in tone so as not to incite further persecution. He drew upon Catholic theologians of the past who had not condemned the heliocentric theory, used other parts of Scripture as proof, and suggested that the judges should be better informed, pointing out Saint Thomas' reply to the impugners of religious orders: "What if the man ignorant of mathematics should attack mathematics, or lacking philosophy, should attack philosophy? Who does not scorn to be laughed at by such mockers?" And Robert Burton, recalling Campanella's defense, extended the charges for English readers by pointing out

> that our modern divines are too severe and rigid against mathematicians, ignorant and peevish in not admitting their true demonstrations and certain observations, that they tyrannize over art, science, and all philosophy, in suppressing their labours (saith Pomponatius), forbidding them to write, to speak a truth, all to maintain their superstition, and for their profit's sake. As for those places of Scripture which oppugn it, they will have spoken for the popular intelligence, and if rightly

> understood, and favourably interpreted, not at all against it. . . . But to avoid these paradoxes of the earth's motion (which the Church of Rome hath lately condemned as heretical . . .) our latter mathematicians have rolled all the stones that may be stirred: and, to solve all appearances and objections, have invented new hypotheses, and fabricated new systems of the world, out of their own Daedalian heads.[21]

He goes on to review the various hypotheses being suggested just as Andreae had presented them in his lectures on mathematics, and Burton whimsically concludes that during all this talk, "the world is tossed in a blanket amongst them, they hoist the earth up and down like a ball, make it stand and go at their pleasures."

Meanwhile, discoveries showed with increasing certainty that the universe operated in regular, predictable ways. It was fast being proved an intricate, mechanistic "clock"—a favorite metaphor of later 17th-century writers. The clock figure appears in Abraham Cowley's College of Prophets, when Mahol, the Reader in Natural Philosophy, is given a special assignment:

> *Mahol* th' inferior worlds fantastick face,
> Through all the turns of *Matters Maze* did trace,
> Great *Natures* well-set *Clock* in pieces took;
> On all the *Springs* and smallest *Wheels* did look
> Of *Life* and *Motion;* and with equal art
> Made up again the *Whole* of ev'ry *Part.*

In no way does this deny God's great powers as Creator; it is a benevolent clock, but His direct intervention and daily attention to the affairs of the universe are no longer necessary when man has mastered His secrets and can put the clock back together again.

With the growing belief in natural law and the effort to determine causes, the supernatural element in religion also received a direct blow. Miracle could often be explained and revelation doubted. Ernest Lee Tuveson, in *Millennium and Utopia,* has shown the transformation of the idea of revelation to fit the scientific age and to support belief in man's progress. "The millennium itself came to be considered as a true utopia—a 'heavenly city of virtuosi,' in which the activities of men even in such fields as science would be enlarged and made more efficient."[22] So progress as a faith and science as its implemental means encroached on the older Christian doctrine of static perfection visualized in the Millennium and in Dante's and Milton's paradise. Scientific activity had at least passed through the pearly gates and was a recognized, legitimate interest of the blessed souls by the end of the century. The new utopists played an active part in this invasion into religion's domain and helped to change the concept of the Millennium.[23]

The great majority of our authors were Protestants living in England, and several were clergymen acutely aware of the conflicting claims of religion and science, which provoked them to express their personal reconciliations. As Englishmen, they did not encounter the organized opposition of Roman Catholicism. The Church of England did not go on record as an opponent, though it did little to encourage the new thought. It was from the rising group of Puritans that the new science received its greatest aid, primarily because it promised practical, utilitarian benefits. Yet, regardless of the particular church, this period of history was only a short 100 years from the beginning of the Reformation, and though the actions of the different churches varied, the old doctrinal tenets were generally used against scientific innovation. Utopists met these religious barriers in their individual ways and, in this instance, their positions represent both the variety of attitudes and the changes occurring chronologically in the 17th century.

To the Protestant Lutheran minister Johann Andreae the Reformation was far from over: he begged his reader with serious intent to help make the world a *Christianopolis*. He shared the energy of the founder of his sect but departed decisively from Luther on the Copernican issue. Luther had reacted as negatively as many Catholic theologians when he called Copernicus the new astrologer and predicted: "The fool will overturn the whole art of astronomy."[24] Andreae's Christianopolitans, it will be recalled, have accepted the new theory; so there is no doubt about their modern outlook. Yet Andreae, like Cowley, Hartlib, and many others, projected a reform that was to be divine as well as human. He saw the entire civilized world living perfectly in this complete unity. In his plans scientific advancement was essential, but one was never allowed to forget religious supremacy.

After extensively praising the attributes of geometry, claiming that it "adapts itself especially to human wants and applies the deepest propositions and theorems to practical matters with admirable diligence," Andreae hastened to add another note showing that he did not advocate a break from religious authority in any way:

> The citizens of Christianopolis, while they measure various things, first of all make an especial effort to measure and weigh themselves, then also to value the goodness of God.[25]

In the midst of the section on arithmetic, which is the "very home of all subtleness," he abruptly reminds the reader that the inhabitants of Christianopolis do not forget the snares of Satan. Though the study of arithmetic and geometry receives emphasis, the mystic numbers of God rank well above worldly mathematics and are learned only by older and wiser men.

> Surely that supreme Architect did not make this mighty mechanism haphazard, but He completed it most wisely by measures, numbers, and proportions, and He added to it the element of time, distinguished by a wonderful harmony. . . . Moreover, these matters are not understood through any human skill, but rest upon revelation and are communicated to the faithful and from one to the other. Therefore they walk into a veritable labyrinth whosoever borrow poles and compasses from human philosophy with which to measure the New Jerusalem.[26]

Religious doctrine generally dominated the lives of Christianopolitans. Printing presses were permitted to publish only the Holy Scriptures and books for the instruction of youth or for the devotion of citizens. Monarchy on the island of Caphar Salama was reserved for God, and the government was administered by a trinity of officials responsible for religion, justice, and learning—in that order of importance. The director of learning made clear to the visitor that the goal was to know God:

> A close examination of the earth would bring about a proper appreciation of the heavens, and when the value of the heavens had been found, there would be a contempt of the earth. . . . Arise, *thou sacred science* which shall explain to us Christ.[27]

Theosophy was the "sacred science" and the highest of advanced studies in the curriculum. "Where nature ends, theosophy begins; it is the last resort, the finding in God what cannot be obtained by physical experiment." In these definitive terms, Andreae limited what the natural philosopher might expect to learn from experiment. The "poles and compasses of human philosophy" would not measure his New Jerusalem.

He felt compelled, however, to reconcile the omnipotent position of religion with his emphasis on science in other parts of the *Christianopolis*, and revealed his discomfort by repetitious statements on religious pre-eminence after enthusiastic descriptions of laboratory research. When, for example, he described the mathematics center, which was "a testimony of human acuteness and energy against mortal chains," and which had complete reproductions of the star-studded heavens, geographic charts of the earth, and every conceivable instrument for study, he claimed that nothing was left to be desired. All mechanical aides were devised to correct and extend the human eye. Yet, he added, "our eyes must turn upward to God, not to the study of stars." Regardless of his uneasiness, Andreae boldly combined his two interests: "For though the sky is so far distant from us and the wings of our original perfection are wanting, *yet we are not willing that anything should take place there without our knowledge*."[28]

Andreae justified man's quest on much the same grounds that Sir Thomas More had used for the Utopians. God intended man to use his noblest faculty, reason. As "divine mysteries impressed upon the land are discovered," they could result only in greater admiration for God's handiwork. No one in Christianopolis denied the value of ascertaining the truth about man and nature. To do otherwise would be to desire the ignorance of barbarians:

> For we have not been sent into this world, even the most splendid theater of God, that as beasts we should merely devour the pastures of the earth; but that we might walk about observing His wonders, distributing His gifts, and valuing His works. For who would believe that the great variety of things, their elegance, advantage, and maturity, and in short, the utility of the earth, had been granted to man for any other reason than for his highest benefit? . . . It is rather man's duty, now that he has all creatures for his use, to give thanks to God. . . . Then he will never look upon this earth without praise to God or advantage to himself; but with an admonition toward moderate use and exact observation.[29]

Andreae struggled with the obstacle of church doctrine, and ended by adopting the solution typical of many other utopists. Christianopolitans continued their harmonious living while pursuing both roads to truth, but with the understanding that different matters were being dealt with. There was no lack of facilities for scientific investigation on a sensible basis and, at the same time, there was no question about God's superior knowledge, which was highly respected, even though they actively delved into it.

In principle, at least, Campanella's citizens of the *City of the Sun* agreed. They saw no animosity in the accepted combination of science and religion and, in the last court of justice, it was understood that the verdict would go to religion as it did in *Christianopolis*. Even in his *Defense of Galileo* against the church fathers who quoted Scriptures to disprove scientific facts, Campanella concluded with submission to the church of his faith. Yet his personal ideas were often at variance with religious authority. Since Solarians accurately reflected his other philosophic writings, his utopia can be drawn upon for his most thoughtful opinions. The three collateral magistrates serving in the *City of the Sun* under the Prince Prelate (Sun or Metaphysician) are Power, Wisdom, and Love—which find their highest, infinite degrees in the Deity and which are limited in every finite creature. In all things that exist these three primal attributes are present and related; thus God moves in all things and errors or sins result from the extent to which one lacks any of these elements. The more remote from God or Being we are, the greater our imperfection and non-being, which express themselves in impotence, ignorance, and hatred (opposites of the first attributes). Solarians understood this and had structured their government, code of ethics, and religion on these three principles that symbolize the Trinity of God. "But they do not distinguish and name the three persons as we do because they are not in possession of revelation." Theirs was a natural religion, the original "hidden" in all men. A revelation therefore was unnecessary; that was needed only for positive religions that might differ and err.[30]

After the Genoese sea captain had explained the Solarians' beliefs, the Hospitaller with whom he was talking comments:

> If these people who follow only the law of nature are so near to Christianity, which adds nothing but the sacraments to the law of nature, I conclude from your report that Christianity is the true law and that once its abuses have been corrected, it will become mistress of the world.[31]

It was a very liberal religious position for the Dominican friar in prison to take, and he showed no hesitation in associating scientific inquiry with religious views. They were not at odds here any more than they were in the actual world of Campanella, if theologians would only study the issues and correct their past mistakes in doctrine.

> Anyone who examines the structure of the world, the anatomies of men (as these people examine the anatomies of those condemned to execution), the anatomies of animals and plants, and the function of every least organ must feel compelled to proclaim loudly the providence of God. Consequently, a man must attend to true religion with deep dedication, and he must honor his maker. No one can do this unless he investigates God's works, attends to sound philosophy, and observes His holy laws.[32]

In the utopian *City*, Solarians suffered no discomfort over the question and proceeded along both roads in an enveloping, Neoplatonic, and mystical unity.

Some utopists, however, went considerably further than Andreae and Campanella in delineating the areas controlled by science and religion. Francis Bacon was first of the group to separate the two decisively, assigning the greater importance in terms of his personal interest and concern to natural philosophy and method. This was the approach that would lead to the new truths man needed to know if he were to control nature and the universe for his own advantage. The problems and methods of advancing knowledge completely engrossed Bacon in all his major writing, and the *New Atlantis* was no exception. He was not a member of the clergy as were Andreae and Campanella. His concern was not with the chain of being leading to God, but with the chain of investigation leading to man's good. And good was not an abstract concept for Bacon, nor was it equated with God as it had been in the earlier humanistic utopias.

The renovation of society in the *City of the Sun* and in *Christianopolis* depended on the spreading of knowledge, science being an important part of this, but magistrates were not chosen merely for their knowledge: religion and moral virtues ranked high as prerequisites for the position. In clear contrast, in the *New Atlantis* "the scientists form a caste endowed with a power superior to that of the King."[33] It can be countered here that Bacon did not include the government and laws for *New Atlantis*, that he intended doing this next, but left the work incomplete. Be that as it may, he chose to write this part first; it was his major interest, and the scientist appears as the unchallenged ruler. The society was Christian and held daily church services, but religion was the underlying bond of unity, a role Bacon had described in one of his early essays: "Religion being the chief band of human society, it is a happy thing when itself is well contained within the true band of unity."[34]

Bacon fully recognized the obstacle of traditional argument to the reception of his *Advancement of Learning*, and began this work by refuting the charge that aspiration for too much knowledge had caused the fall of man and that the search for knowledge may lead to a swelling of man's pride. It was not too much knowledge, but the presumptuous desire to penetrate the Divine mind and become like God. With customary thoroughness, he answered the opponents' claim and presented the case in favor of the questioning process as completely consistent with Christian principles. Natural philosophy was simply and clearly separated from theology. Man could not hope to know God's nature or inner mind by employing the reason and sense of Bacon's method. His Supreme Being could be understood only through revelation from the inspired Word and faith. The human creature can only aspire to know what is set forth in His works, and for Bacon this was more than sufficient as the goal. He made this distinction in his early, middle, and later writings; he had no doubt. And his definition of metaphysics made it even clearer: metaphysics was a generalized physics, the summary laws of nature that had no connection with theology, no concern with anything beyond nature's laws. It was far from Aristotle's First Philosophy. When Bacon published the *Novum Organum* (1620), he decisively stated his point and severed the two fields of thought. Matters of religion were not permitted in his laboratories.

> The corruption of philosophy by superstition and an admixture of theology is far more widely spread, and does the greatest harm, whether to entire systems or to their parts. . . . Yet in this vanity some of the moderns have with extreme levity indulged so far as to attempt to found a system of natural philosophy on the first chapter of Genesis, on the book of Job, and other parts of the sacred writings; seeking for the dead among the living: which also makes the inhibition and repression of it the more important, because from *this unwholesome mixture of things human and divine* there arises not only a fantastic philosophy but also an heretical religion. Very meet it is therefore that we be sober-minded, and *give to faith that only which is faith's.*[35]

It was this type of statement that led to the charge of Bacon's being antireligious, even an atheist.[36] But he was no atheist; he decided upon a separation of the two areas and, in addition, he was simply less interested in theology than in natural philosophy. He considered religious dogma a handicap to the procedure he advocated as the method of learning. Things human and divine were not to be mixed. His zeal was for the Kingdom of Man rather than for the Kingdom of God or Heaven. Bacon's sermon in the *New Atlantis* showed his deep belief in his scientific method. The Father and members of Salomon's House were the most revered citizens of Bensalem. To a member was given the ark containing the teachings of Christianity. It was the Father who had "an aspect as if he pitied men," and who held up his hand as if in blessing. He appeared almost as a surrogate for religious leadership in the community, though Bacon probably did not intend such extreme symbolism. Much earlier, in his *Refutation of Philosophies* (1608), he had described a philosopher delivering an imaginary address before an assembly of sages. Here too, the speaker had "an aspect calm and serene, but habituated to an expression of pity."[37] Apparently Bacon's mental image of the wise philosopher remained as consistent as his thought and purpose.

The only miracle that ever occurred in Bensalem accompanied the introduction of Christianity. Yet even on that occasion the member of the scientific college who uttered the prayer was careful to maintain his superior knowledge by distinguishing between a miracle and an imposture. Bacon had never categorically denied the existence of marvels or miracles any more than he completely repudiated the possibility—by chance—of good to come from alchemy and natural magic, though these pursuits were fraught with deception and imagination rather than honest endeavor and reason. In the *Advancement of Learning* (2.3) he had considered marvels and miracles worthy of analysis and study that might bring forth a germ of truth at least. To King James I he had written commending His Majesty for looking "deeply and wisely" into such shadows as these "with the two clear eyes of religion and natural philosophy." And he had decided concerning miracles of religion that "they are either not true, or not natural;" in either case they were impertinent to the study of nature. So the miraculous entrance of Christianity (the Bible and other documents were found encased in an ark on the water when a great pillar of light with a large cross on the top had dispersed as stars into the firmament) is no evidence of confusion in Bacon's mind nor in the opinion of Bensalemites, for the representative of the scientific brotherhood who prayed at the great ceremony spoke like a man who had Bacon's legal experience as well as his scientific interest in determining the true from the false:

> Thou hast vouchsafed of thy grace to those of our order, to know thy works of creation, and the secrets of them; and to discern (as far as appertaineth to the generations of men) between divine miracles, works of nature, works of art, and impostures and illusions of all sorts. I do here acknowledge and testify before this people, that the thing which we now see before our eyes is thy Finger and a true Miracle.[38]

This miracle simply was unnatural and belonged to religion. Furthermore, God had directly granted the scientific foundation the right to discern the difference and to know the secrets of His six days of creation, for which purpose Salomon's House existed. But once the secrets were discovered, another world was to be created by man, a new and better kingdom for man. Bacon's optimism set no limits for man's knowledge such as the ones Andreae imposed in his constant reminders of God's omnipotence.

The Lord Chancellor, by the end of the first quarter of the century, had solved the conflict between religion and science to his own satisfaction by separating their areas of authority. This was not to undervalue religious truth and meaning; it was merely to clarify the purpose and approach to both types of knowledge. Religion remained in the scientific utopia of *New Atlantis* as a functional, unifying basis for society and the original sanction for the distinctive work of Salomon's House. Once again, the "illustrious Lord Verulam" set both the tone and the pattern for many utopists of the next four decades. The tone varied somewhat with the individual author's interests, but the pattern of the new utopia was fairly well established insofar as its focus was man, not God; this world, not the one to come either in heaven or the Millennium.

While Samuel Hartlib's Macarians lacked Bacon's distinct separation between faith and science, they had generally adopted Baconian attitudes and taken a definite position regarding man's ability to bring about the reformation of the world himself. When the scholar in Hartlib's dialogue comments that many "divines are of opinion that no such reformation, as we would have, shall come before the day of Judgment," the traveler, who has seen Macaria, replies that a hundred scriptural texts plainly prove them wrong. Saint Jerome is correct in saying that though one may expound the allegorical Scriptures in a spiritual sense, "the prudent reader will by no means receive such an exposition." The two gentlemen agree that they must be "good instruments" in the task of remaking the world, which has "natural causes" to aid it. Macaria is an improved Christian society in which natural philosophy and its application are firmly established. The new order of clergy is trained in medicine. Moreover, the work of experimenters is in no way hindered by the old guard of divines who pessimistically assert that not before the final day will the reformation come to pass.

Among the new utopists who reacted to religion's authority, Gerrard Winstanley expressed the most extreme and violent opinions about the church.[39] As he fiercely berated the clergy for deceiving the people, he denied the Resurrection and a spatial heaven and hell. Heaven was to be sought in this world, and salvation found in liberty and peace. The means he advocated were economic reform and advancement of knowledge—certainly not the clergy's promises, which were only to obscure the devilish purpose of keeping men ignorant in order to exploit them. He would abandon the clerical hierarchy and all outward forms of religion.

Winstanley did not reject God or Christianity as a faith, but he demanded a most intelligent, rational God. And in several passages he questioned the possibility of a life hereafter:

> For . . . to know what he will be to a man after the man is dead, if any otherwise, than to scatter him into his Essences of fire, water, earth and air, of which he is compounded, is a knowledge beyond the line or capacity of man to attain to while he lives in his compounded body.[40]

The people of his New Israel were not to expend their energy or hopes on such chimeras as the world to come. Instead, he envisaged a brotherhood of man and the enjoyment of earthly freedom by all. His holy commonwealth was preparing the way for God's action, which would bring the perfect life into being in this world.

He was a bold extremist and left no doubt where he stood on the issue, while others like Petty and R. H., on the opposite end of the scale of reaction, paid little or no attention to the relationship between science and church doctrine in their utopias. Increasingly, interest focused on the more exciting scientific developments, and later in the century the religious opponents were neither so powerful nor so talkative. Elsewhere in unpublished papers, however, Petty expressed himself at length in favor of uniting all Christians on common beliefs and, more important, he spoke for natural religion and plain, understandable doctrine. Reason was used not to support religion, but to criticize it. As Richard Westfall in his comprehensive study of religion and science in the 17th century has pointed out, this was an extremely vital change after which natural religion began to replace rather than uphold Christianity.[41] It could be sufficient to worship God from the evidence in nature alone, and already the virtuosi regarded nature, the object of their study, with awe and fascination.

The last utopian statement of Bensalem fittingly summarized religion's role in that community. Arguments of utopian societies and colleges, published in the 50-year interval between Bacon's first description of *New Atlantis* and Glanvill's "Continuation," had telling effects on the opinions of those living in Bensalem. Now it was a wholly rational society, not at all susceptible to belief in the church doctrines that discouraged and opposed the new science at the beginning of the century. Divines in the Bensalem of 1676 were no longer influenced by religious theories of the schoolmen, but thought

> that those subtile, and Angelical Doctors have done Religion no small disservice, by the numerous disputes, niceties, and dis-

> tinctions they have rais'd, about things, otherwise plain enough. . . . And they judg'd, There was less cause in the latter ages to reckon of School-Divinity, since the Peripatetick Philosophy, on which it was grounded, grew everywhere into discredit: So that they thought it not safe, to have Religion concern'd, in that, which did not truly help it; and which was not now able to help itself.[42]

The new facts of science had thoroughly undermined the Ptolemaic and Aristotelian universe. Scholastic divinity could no longer help itself, much less contribute to the solution of current problems of religion. " 'Twas time now, in such an Age as *this*, to assert the sober *use of Reason*, and to rescue Religion by it." The church of Bensalem had not only been rescued but reestablished firmly on the foundation of reason, "clear and distinct" thought, and liberty of inquiry. Glanvill's basic tenet—like Campanella's and Bacon's but for different reasons—was that attempting to discover the secrets of nature was in no sense inimical to religion. His God, like Winstanley's, was entirely rational and would never consider squaring a circle; to act arbitrarily was an imperfection in the Almighty Being. This question had long been discussed, but now it was used in a new argument. Consistent with this attribute of God was the emphasis on man's use of reason, his greatest gift.

To be sure, Glanvill said that there were areas in which man's ignorance was abysmal; for example, man could not tell "how we speak a Word, or move a Finger; How the Soul is united to the Body." The narrowness of human capacity, however, was neither cause for discouragement nor final evidence of man's inability to understand God's secrets as it had been for Samuel Gott in *Nova Solyma*. Glanvill explained in this context that Bensalemites, aware of their ignorance and their proclivity to error, were simply more skeptical, more reserved in affirming or accepting a theory. They realized the possibility of being wrong and "there is something then to work upon towards their better Information."

Divines were well schooled in natural philosophy, particularly in "the right method" of thought, which enabled them to consider theories clearly and then test them for proof to the senses. With this in mind, it was not surprising to find ministers believing that God's grace was not given only to the few elect, but operated as a "general Cause . . . in conjunction, as the Sun, and moysture of the earth, and seminal principles do in the production of Plants and Flowers; each cause doing what is proper to it." Free grace became a natural process. The naturalistic explanation was also given by educated churchmen to discredit fanatics—the religious enthusiasts who claimed their illuminations and prophetic raptures were divine communications. Such ecstasies were only a natural disease—far from sacred or supernatural—and could result from a strong fancy working on violent affections. Imagination was capable of extraordinary apparitions, voices, and revelations, so that a zealot thought himself a prophet and by his vehemence caused others to believe. Such deceptive dreams were thoroughly exploded in Bensalem by the use of cool, deliberate reasoning. A similar reasonable analysis operated on matters of moral philosophy, for the old "Disputing Ethicks" had been abandoned. Bensalemites established their rules for behavior upon the study of human nature as it truly is. They observed the inclinations, humors, and appetites of man himself, and then constructed their system of ethics.

In Glanvill's community were no "aery notions" flying about to deceive men. Reason had permeated the very structure of the sermons delivered by these clear-headed divines. Their preaching was practical and direct, with no show of ostentation or appeal to emotions through hypnotizing rhetoric. Only plain texts and early church writings (before A.D. 500) were used—not the "dark places of *Daniel*, and the *Revelations*." The same orderly division of thought necessary to advance any knowledge was applied to the organization of material for the sermon. Ministers

> divided their matter into the *substantial* parts of Discourse; or resolv'd it into some main *Proposition;* and so treated of their subject in the method that was *natural* to it, and most beneficial for the people they were to instruct. . . . They did not talk by roat out of Books, or *Enthusiastick experiences;* they did not direct by *Metaphors*, and *Phrases*, and *unpracticable* fan-

cies: But laid down the true, sober, rational experimental method of action.[43]

Glanvill's emphasis was seen in his very choice of words: matter, substantial parts, proposition, method, natural, beneficial, and so on. Furthermore, his selection of topics and attitudes toward them were characteristic not only of the climate of thought in utopia, but also of the intellectual atmosphere in England which was ready to entertain the theories of John Locke. As Basil Willey has suggested, Glanvill was not a highly original thinker, but he "reflects in miniature the transition from the earlier to the later phases of the century."[44]

Glanvill was an active member of the Royal Society when he was Chaplain in Ordinary to His Majesty. These two areas of interest were combined most effectively and easily in his continuation of the *New Atlantis*. His solution, however, was not based on Bacon's separation of the two roads to truth. Glanvill applied methods used in scientific advancement to the analysis of religious matters. Neither area of thought was to rest on uncertainties. Both were to be founded on infallible truth that could be proved to the senses and to reason. His was one road in method, and the facts discovered were equally valid and compatible.

The utopian society had changed with the birth of new worlds of thought, and Bensalem had now entered the age of reason and the age of tolerance. No longer would religious doctrine deter the progress of knowledge. The utopists' echelon stood firmly in favor of a God who was honored by their efforts to learn and to use their reason. They were resolutely opposed to Aristotelianism and the outmoded curricula of established educational institutions together with the crippling of energy by undue worship for the ancients. The new utopists thus established their solid positions against the three authorities that had defined man's beliefs and enchained his mind.

Once the principle of an "open mind" was established, and the critical attitude of rational skepticism applied, modern science was ready to find its new methods and rapidly move ahead.

With no doubt whatsoever in the new utopists' minds, the key to the brilliant future was the method that was to unlock nature's door hitherto closed and reveal the shining gems of knowledge never before known. Moreover, when they considered the ways and means to accomplish this miraculous task, they unhesitatingly endorsed what Joseph Glanvill called the "*right Method* of Studies" which "would go very far, and do mighty Matters in an indifferent Time."

VII

THE RIGHT METHOD

THE RIGHT METHOD of Glanvill and the new utopists as a group was the "*mighty Design, groundedly laid, wisely exprest,* and *happily recommended* by the *Glorious Author*," who was Francis Bacon. The italics are all Glanvill's. This tribute, however, referred only to the Lord Chancellor's advocacy of the inductive process, the method of empiricism based on collection, observation, reason, and experiment. Aside from this general commitment, the scientifically minded utopists were far from a clear definition of method. Bacon had sold the idea through his great prose style—his simplicity and clarity of expression were highly effective, with common, memorable metaphor—and his well-organized approach to a great, all-inclusive endeavor, his *Instauratio Magna*, which was a sufficient challenge to any man interested in reorganizing and advancing knowledge. Bacon presented the method in detailed steps, exact and limiting compared to those before him, but he still left ample room for others to fill in and add their own notions. This Glanvill certainly did. Although he gave the glorious Bacon credit for creating the "mighty Design," his own process differed markedly. By the time he projected *New Atlantis* in 1676, Descartes, Henry More, Hobbes, Gassendi, and Mersenne had all contributed to this question; so the later Bensalemites had thoughts regarding method that were quite different and considerably advanced when compared with their founding father's.

Foremost as a goal for the new method, regardless of the individual author's bent, was to find out *how* things worked, the cause that would produce predictable effects, the chain of consequences. As they had readjusted their concept of God and church doctrine to fit the new goals, so also they no longer asked the *why* of actions, they no longer saw the universe as purposive in the sense of an object seeking to fulfill itself in a higher, more perfect function. Instead, the object would act according to its own nature and the forces acting upon it: a mechanistic operation, inevitable, and in accordance with natural laws. Hence, there was an about turn in the direction of thought and energy, away from the teleological universe of the Middle Ages in which all things took their assigned places according to a preconceived and perfect structure. Metaphysicians shifted emphasis from "meta" (meaning beyond or transcendental) and began to concentrate on its other meaning of next in place after physics, which promised better answers. Aristotle had so defined the word, but for him the next things to consider after physics were First and Final Causes, Pure Form, and the Prime Mover. This Bacon changed: knowledge was a pyramid based on natural history and ascending through physics to metaphysics, which pronounced only the highest natural laws. Ontological considerations were out, and he proposed a thoroughgoing materialistic naturalism. As early as 1609 in his work *De Sapientia Veterum*, Bacon had commented that "controversies of religion must hinder the advancement of science." Agreeing with this, members of new scientific academies like the Royal Society in England and the Academie Française did not permit themselves to discuss theology or philosophic subjects that might lead the work off into vague considerations. Subjective and qualitative approaches gave way before the disinterested and objective method that increasingly depended on mathematical and quantitative measurement. Discussion on the "essence" of matter was replaced by trial of the senses and the experimental process. There was less concern with the microcosm, man, as he reflected the great order, and more concentration on man as the means with the power to measure.

The utopists' problems were further complicated because they set for themselves an insuperable task: agreeing with Bacon's goal, they expansively embraced the job of assembling the whole of learning, facts, and experience. Pansophic colleges like Comenius' and compilations of "natural histories" to record every known fact were advocated with complete seriousness as absolutely realistic goals. In the *City of the Sun*, a man educated in only one area of learning was considered unlearned and unskilled. This enlightened city counted ignorance a sin, and the culprit was punished by forced instruction in those sciences and arts that he had neglected. At Bensalem and in Andreae's Caphar Salama, experiments were designed to cover every conceivable possibility and to yield results for the grand total of *all* knowledge.

Theories of advancement in science and learning were founded on the idea that knowledge was finite and a unified whole which one mind could embrace. Once the basis of knowledge was established and the proper method for distilling it discovered, all knowledge could be advanced by the same method. The problem was fundamentally epistemological. We saw evidence of this in the way utopists often moved the laboratory method into the classroom and vice versa. In the excitement of the new science it could hardly be expected that methods would be differentiated according to the nature of the object or goal set.

A letter of Samuel Hartlib, written in 1630 to his friend John Dury, illustrates both the general confusion regarding methodology and the concern over its importance. Hartlib questions:

> Whether hee were best to bee taught the fullest and best ordered systemes which as yet wee can have, and that tabularized. I meane the whole method of an Art or Science with all the definitions, divisions and canons, together with the feat of Ars Universalis. Or rather to follow my Lord Verulams directions in his so much commended Aphorismes as the onliest way for deliverie of knowledge. . . . For rejecting the former systematical faggotting of precepts into a sensible method, hee chooses the course of Aphorismes, which except they should bee ridiculous and unservicable . . . cannot bee made but of the pyth and heart of sciences. For discourse of illustration . . . of connexion and order, descriptions of practice and the like must be cut off; and nothing remaine to fill the bodie, but

> some good quantitie of orderly observations. . . . Aphorismes representing a knowledge broken doe invite men to inquire further. Or thirdly whether the maine principles or seeds of all Arts and Sciences only should bee proposed, with a compleat After-Art of drawing consequences, which might bee profitably thought upon. Or lastly your more accurate and new Analytical and Genetical way, together with that of Mr. Reineri and Acontius.[1]

With this letter a rude draft of John Pell's method was enclosed, which, G. H. Turnbull thinks, was probably his *Idea of Mathematics*.[2]

Although Hartlib clearly preferred the Baconian method, the problem had obviously not yet been solved in his mind, and he was not alone in his ambivalence. Not until the 18th and 19th centuries did scientists generally feel complacent about their methods and withdraw from philosophic speculation concerning the implications of new facts discovered in the laboratory.[3] During the 17th century, the new scientific approach was consciously creating a new philosophy of the universe and man's place in it. The question of scientific method was, therefore, central from every point of view. And, in spite of the vagueness, various approaches began to emerge, though the categories were not mutually exclusive. The papers enclosed in Hartlib's envelope indicated the alternative methods under discussion: the mathematical approach, the Aristotelian method of classification, and Bacon's type of empiricism. After considering these three possibilities, the utopists decided in favor of a nonmathematical, antideductive, experimental method.

Nonmathematical Approach

Recent progress in mathematics had been remarkable, making it a major factor in the advancement of science,[4] yet only a few utopists realized its importance or its ability to convert facts into "laws." While scientists like Galileo and Kepler were employing quantitative methods and mathematical analyses in their explorations, the utopists—literary scientists—generally advocated the empirical method without any real notion of the part mathematics was to play in it.

The outstanding exception to this was Andreae's *Christianopolis*, in which mathematics was heralded as a vital means to penetrate the darkness of human ignorance. He alone among utopists correctly anticipated the quantitative method, which was destined to reveal great wonders of the universe. Both in the laboratory and in the educational curriculum of *Christianopolis*, the power of mathematics was the subject of eulogy:

> If you consider human need there is no branch of knowledge to which this does not bear some help of first importance. If you consider the undertakings of man's mind, you will discover that man struggles almost with infinity, in this one direction, and worms his way far into the secrets of progression. . . . Hence, this study is pursued by the inhabitants of Christianopolis with the greatest perseverance, and every day they find in it something to admire, something which sharpens their wits and lessens their labors. In algebra they have no equals, because it calls forth all the powers of man, treats physical units in an entirely unique manner, and solves the most intricate problems with incredible keenness.[5]

Geometry has almost magical powers. It measures not only regular shapes, not only dimensions that are near at hand, but all figures.

> It passes through them, changes, balances, transfers, raises and plays a most elegant part in all human labors. If one desires theoretical research, nothing is more subtle; if one desires to apply practical problems, nothing is more convenient or rapid. If you intrust to it any talent, the same is returned nimble and applicable to anything. Hence, the inhabitants of Christianopolis set much store by it, since they see that there is no art which is not rendered easier by it, and that man becomes more expert for taking up such arts.[6]

In contrast to the wisdom of Christianopolitans, Andreae bewailed the fact that the outside world failed to see the immense importance of mathematics, thus casting aside half of learning:

> Therefore, until those who profess to be broadly educated without mathematics shall return into her favor . . . I will pronounce them only half-educated, and they shall bear testimony to this accusation against themselves whenever they shall suffer themselves to be led forth upon the forum of human sciences.[7]

Of the six laboratories in this ideal community, two were devoted to mathematics and mathematical instruments, which were extolled for their values to the burgeoning study of astronomy.

Andreae's happy wedding of astronomy and mathematics and often his words on measure and calculation have the ring of Plato's phrases on the same subject. Plato, too, had advocated extensive study of mathematics, but for him it was a discipline needed by the fathers of his *Republic* to sharpen their reasoning powers for contemplation of the abstract; it was an introduction to philosophy. For this same reason, over the entrance of his famous Academy the inscription read "Let none who has not learnt geometry enter here." And those who entered found mathematics in the curriculum, and found themselves with the assignment of seeking out rules by which "the movements of the heavenly bodies could be reduced to a system of circles and spheres."[8] In his *Timaeus*, with all its Pythagorean overtones and clouds, Plato mathematically explained the universe, and in the *Theaetetus* (143b) he described his concept of God as a geometer. He thus asserted belief in a logical system and suggested a pattern of relationship between mathematics and astronomy that endured 2000 years, to the time of Kepler; but to Plato these various uses for mathematics were all theoretical and explanatory, not exploratory. It was a symbolic use in the highest sense and reserved for dealing with unchangeable, eternal verities as he saw them; it was an "initiation into the mind of God."[9] There was no connection with material things, with the senses, with experience, or with observational activities. Training in mathematics was for the study of pure forms—the quest for knowledge of absolute virtue—which was philosophy and theology for Plato.

Certainly Andreae could believe in this Platonic doctrine generally, but he could not stop there: he applied mathematics to all areas of human knowledge, and regarded it as a short cut, a labor-saving device in both practical and theoretical matters. He did not repudiate with Plato the material object and the power of the senses; instead, his mathematics entered the laboratory.

Campanella, too, sounded Platonic with the Pythagorean mixture upon occasion. In his hierarchical universe he differentiated between the "eternal or mathematical world and the temporal or corporeal world."[10] And, without detailed explanation, he placed mathematics on the first wall of learning in the *City of the Sun,* merely explaining that here "all the mathematical figures—more than Euclid or Archimedes speaks of—are shown in their relative proportions." Although he gave it this primary position and mentioned it as an important area of knowledge for all citizens as well as the high ruler, he said nothing of its specific attributes. Similarly, Glanvill, in the last report we have of the life of Bensalemites, stated they were "great valuers" of mathematics, "which they accounted excellent preparatives and helps to all sorts of Knowledge." And such study, he believed with Plato, had the special advantage of training the mind "to a close way of reasoning;" it was thus a good antidote for the confused meandering of contentious schoolmen. Some Bensalemites think that for this reason mathematics should be the basis of training for youth. Elsewhere in his writings, Glanvill had echoed Hobbes, praising the study as "the indisputable Mathematicks, the only Science Heaven hath yet vouchsaf't Humanity."[11] It was a broad endorsement, not qualified in terms of specific values to natural science, but in this statement Glanvill summarized the general position for the utopian writers.

William Petty had seen the applicability of mathematics in the assessment of social and economic problems, but he had not extolled its use in the advancement of scientific knowledge itself. Cowley designed a "Mathematical Chamber furnisht with all sorts of Mathematical Instruments," which he annexed to the library in his experimental college. He assigned one of his 16 professors to teach this study, so the boys learned geometry together with astronomy. No further comment was made; it was, in effect, general recognition of the field, no more. And Bacon, frequently criticized for his failure to appreciate the powers of mathematics, neither assigned it a prominent place in his process nor recognized it as an especially important area for study in the *New Atlantis*. There is one brief passage in which the Father of Salomon's House claims: "We have also a mathematical house, where are represented all instruments, as well of geometry as astronomy, exquisitely made."

In Bacon's defense, it has been pointed out that he realized that "many parts of nature can neither be invented—that is, observed—with sufficient subtlety, nor demonstrated with sufficient perspicuity . . . without the aid and intervening of the mathematics."[12] Bacon recognized as valuable the Pythagoreans' opinion that numbers could penetrate the principles of nature, since quantity when applied to matter was causative of many effects in natural things, but he felt that the theory of numbers, like Aristotle's logic, had got out of hand and was exercising dominion over physics. Mathematics had a way of departing from facts and constructing notional systems of its own. Instead, it should be an auxiliary science, a handmaid to physics. With this position assigned in his *Novum Organum* (2.44), Bacon included among Mathematical Instances its powers to measure time, quantity, relative strengths, weights, and distances. Thus he employed mathematics as an aid though he did not stress the study excessively in his method or its use in his utopia. Such statements are not frequent in Bacon's works, which usually concentrate on clearance of the erroneous notions that beset men's minds and advocate the new way as he saw it. In the *De Augmentis* (1623), which was an enlarged Latin version, with some discreet deletions, of the *Advancement of Learning*, Bacon claimed that in arithmetic there was still lacking "a sufficient variety of short and commodious methods of calculation, especially with regard to progressions, whose use in physics is very considerable." This he wrote nine years after Napier's logarithms had appeared and been greeted with enthusiasm by Kepler, who introduced them into Germany.[13] And yet Bacon was right: Descartes, Leibniz, and Newton were soon to add better methods and greater concepts. No new developments were enough to satisfy Bacon's insatiable desire, his "hydroptique thirst," as it would have been called in his day. Even as he criticized Galileo and Gilbert for an inadequate number of examples tested, so he saw the need for discoveries in mathematics, in instrumental devices, and techniques to aid research. This was one of Bacon's strongest beliefs that spread to others and stimulated them to new ideas and work.

If Bacon somewhat underestimated the value of mathematics, his successor, the anonymous R. H., remedied the situation for him and gave it at least a nominal position of prominence. In the great hall of fame through which the visitor is led by Joabin, who had guided Bacon's tour of *New Atlantis* 33 years earlier, four mathematicians are included. Here on one side of the long gallery we see

> a statue of *J. Neper* Baron of *Merchiston* who first invented the whole use of *Logarithmes.* And next to him were erected the statues of *Johannes Regiomontanus,* who made the wooden Eagle and iron flie; and *Erasmus Rheinhold,* who transcended all in the rules of Tangents and Secants. And not far from these he signally pointed out the statue of that most learned Geometrician *Thomas Harriot,* who was the first, he told me, that found out the *Quadrature* of the Circle.[14]

After the statues were identified, there was no further discussion of the work of these men, and R. H.'s attention rapidly shifted to the inventions that so captivated his imagination. Yet he was second only to Andreae in stressing the gifts of mathematics. The others merely mentioned it in passing and none (except Andreae) evinced a strong belief in the power of mathematics to advance knowledge. Instead, the utopists as a group belonged to the nonmathematical current of scientific development.

Antideductive Attitude

The utopists' general lack of foresight in recognizing the power mathematics was to have in scientific advancement may have been due to chance—oversight—or ignorance of current developments in such a specialized study, but it may also have resulted from their psychological distrust of symbolism: as words were "creatures of the brain" rather than "things" to see and touch, so numbers were concepts replacing the concrete object. They would have no operations, even one step removed, from the actual object. This attitude was clearly seen in their reforms recommended for the curriculum and in their repudiation of words and dependence on ancient authorities embalmed in textbooks to be memorized and regurgitated. Just as this routine was expelled from the classroom, so was it banished from the laboratory. Utopians no longer educated or respected man as a logician who was apt to forget his vital role as an observer of nature.

Amid the complex of notions being abandoned, the teachings of Aristotle continued to be a major issue. To him was attributed fatherhood of the deductive method, which spun words from untested hypotheses into entangling webs of thought to the point of no return and certainly no new knowledge. Because of the relative novelty[15] of the experimental approach that the utopists vigorously supported, they felt compelled to argue its merits against the rationalism of the deductive method. When they shouted fanatically against reason and words, they were understandably extreme in their effort to defeat finally the opponent, the schoolman who was "the Ghost of the Stagirite, in a body of condensed Air," and who had overused and misused the rational approach, reducing it to a game of verbal deception.

Hartlib, in his letter concerning possible methods to be used in learning, had rejected the "systematical faggotting of precepts" and "discourse of illustration" together with descriptions of the qualities of natural phenomena. He was directly castigating the major features of Aristotle's method of classification and syllogistic logic, which he considered inferior to Bacon's "Aphorismes" as a way to further inquiry. So, too, Glanvill complained that Aristotle explains nothing, he only restates, and eventually says "this is so because it is so."[16] And he concluded that following Aristotelian procedures would never reach "the Treasures on the other side of the *Atlantick*," which were the new facts to be discovered.

Any one of the utopists might have spoken these words. On this all agreed, though some were less specific in their charges, hitting broadside at dependency on reason and Aristotle as its exponent. They were not the first to throw stones in this direction, of course: Telesio, when proposing his correction for Aristotle's concept of the relationship of matter and form, had roundly berated the dominant ancient,

> Aristotle, whom now for so many centuries the whole race of men venerates like a deity, and, as though he were taught by God Himself and the interpreter of very God, hears with supreme admiration and even with supreme religious observance.[17]

And he had gone on to plead for the examination of nature herself, lying ready for investigation if men would but proceed. This his disciple, Campanella, was ready to do, stating his delight in Telesio's "liberty of philosophy" and his dependence on "the nature of things, not on the sayings of men." And Campanella added the old religious grounds for repudiation: Aristotle was a heathen. Campanella wished his scientific new world to be firmly, if mystically, in accord with Christianity.

The same issue beclouded his theory of knowledge, which was closely integrated with his complete and religious concept of the universe. He asserted that we each possess a hidden self-knowledge that we have to learn about from our acts and from external influences that cause continual modification. This self-knowledge is the necessary presupposition for knowledge of all other things, which springs then from two sources: sensible experience and reasoning. For endorsement of this theory Campanella turned to Augustine's claim for immediate consciousness: "As for me, the most certain of all things is that I exist." It was an early statement, repeated by several philosophers of the Renaissance, that culminated in Descartes' theory in his *Discours sur la méthode,* which appeared one year before Campanella's publication, but the latter had written his exposition largely while in prison years earlier.[18] In any case, Campanella departed abruptly from Descartes in clarity of presentation, terminology suitable to the scientific thought, and in the final conclusions.

Campanella remained a child of the metaphysical world in which he grew up, and he philosophized in the scholastic fashion, deductively, though he condemned the approach in the study of natural phenomena. Here he spoke most clearly for proof to the senses and above all senses, the visual. The walls of his *City* were built of knowledge even as the walls of Andreae's laboratories in *Christianopolis* were decorated with the phenomena of nature in action, for "it is all of no value unless you shall see everything before your eyes."

When it had focus, the utopists' attack was against *a priori* speculation. It was rebellion against tricky systems of logic that led to the excommunication of unaided reason as a means for obtaining knowledge. Reason was doubtless man's highest faculty, as Abraham Cowley was frank to admit, but reason alone could "create nothing but either Deformed Monsters, or at best pretty but impossible Mermaids."[19] The rational element, of course, continued to play its inevitable role in the evaluation of facts, the determination of relationships, and the formation of conclusions. But for the utopists, the scientific process had to start with the visual element, build upward through reason to the truth, which was finally confirmed by observation. It was Aristotle's failure to perform in this way that drew Bacon's censure. Among those who "wrest and corrupt" philosophy by their "preconceived fancies," it is Aristotle who "affords us a signal instance." He "made his natural philosophy completely subservient to his logic, and thus rendered it little more than useless and disputatious."

And Glanvill qualified his disapproval of peripatetic philosophy by saying that its prominence was inordinate, that it had drowned other more valuable approaches. His was not then a wholesale rejection of classical authorities any more than Bacon's was. Both would examine carefully the older theorists to see if truth lurked beneath or around their words. And both came forth with admiration for Democritus, whose theories returned to the favorable light in the 17th century as we saw in the attitude of Robert Burton. The atomistic philosopher with his materialistic view of matter fitted the developing modern theories and the increasingly secular tone of writings on natural philosophy. Furthermore, Democritus had observed and anatomized, so Bacon lauded his attempt to start from the particular:

> It is better to dissect than abstract matter; such was the method employed by the school of Democritus, which made greater progress in penetrating nature than the rest. It is best to consider matter, its conformation, and the changes of that conformation, its own action, and the law of this action or motion; for forms are a mere fiction of the human mind, unless you will call the laws of action by that name.[20]

Yet, a few passages later in the first book of the *Novum Organum,* Bacon in his characteristic way of thinking cautioned against this approach, too, if overindulged. When comparing those who were distracted and weakened by study of the particular, individual form with those who were stupefied by contemplation of the whole structure, he cites, as his example of the former flaw, Leucippus and Democritus, "for they applied themselves so much to particulars as almost to neglect the general structure of things." Bacon would have both, and his method was the compass to guide man safely and steadily toward the whole truth, which was based on the amalgamation of particulars.

Bacon's use of the word "form" immediately raises confusion concerning what he meant. He used the term neither in its scholastic sense nor in its Aristotelian meaning; he redefined it, as often was his custom with words, for his own use. And he equated it with law. According to Bacon, form is the primary quality of a body, the substance or essential nature, which stands in the relation of cause and effect to the secondary qualities or attributes. When the form is determined, it has the power of a law. Form is both the basic constituents of a thing and the law of its activity; hence forms constitute the alphabet of nature. His definition is a step away from the metaphysical view in the direction of physical science (just as Galileo distinguished between primary and secondary qualities but with different results). Bacon intended his inductive method to concentrate on defining this form, the primary qualities of a body, which he considered to be the "abstract" nature upon which other qualities were dependent. Then he included lists of concrete bodies for study that confounded the issue. At least he thought further on these matters than other utopists and defined his position from his own point of view.

For example, when comparing his new "ascending" process with the older induction used by logicians (which he found wanting, even "utterly vicious and incompetent," although at one point he acknowledged a "swimming anticipation" of his method in both Plato and Aristotle), he drew the lines of difference warily and thoughtfully. Whereas the older induction merely enumerated particular cases and tended toward useless generalities, his method was new in its principle of rejection or exclusion, in its inclusion of the contradictory, and in the number of examples he would collect to attain certainty. Most important, his process was to start with man's mind clear of notions, a *tabula rasa* as it were. So, too, Bacon was prudent in dealing with the deductive process. He did not wholly rule it out; instead it was a question of when and how it had value. Consequently, he would allow the "descent" in reasoning when the results of induction were to be practically applied. He called it the "correlative process of deduction." Thus, while the utopists indicated general dissatisfaction with the syllogistic or deductive method and Bacon joined the chorus, he wisely examined the role of reason and assigned a suitable place to the deductive approach.

After the Aristotelian lion of rationalism had been bearded, a writer could feel easier about the challenge and become more "reasonable" himself. Thus, as noted earlier, toward the end of the century Glanvill's skepticism permitted citizens of *New Atlantis* to consider calmly the various theories and their merits. But all utopians were through with the shrouds, the winding sheets, of rhetoric: the new life was to be free from all such encumbrances and gloomy restrictions. Facts were to be looked at squarely and clearly, then accepted as true or false on the unquestionable basis of trial and experiment. Glanvill's utopians examined "with *freedom* and *indifference*" and concluded according to the "Report of their Faculties." Gerrard Winstanley's utopians also recognized truth only from the crucible of experience. When a man rose to speak on the public forum in utopia, he was "required to speak nothing by imagination, but what he hath found out by his own industry and observation in tryal." This was the recurrent theme of all new utopias. And for the experimental approach the utopists were as dogmatic as their predecessors had been for the deductive.

Experimental Method

This was the *right method*. But while the scientific utopists aggressively supported firsthand observation and the collection of facts, they sketched only vaguely the precise steps for following this new procedure. Frequently their suggestions remind one of Comenius' so-called "specific" method of the sciences set forth in his *Great Didactic:*

> Science, or the knowledge of nature, consists of an internal perception, and needs the same essentials as the external perception, namely the eye, an object, and light. If these be given, perception will follow. The eye of the inner perception is the mind or the understanding, the object is all that lies within or without our apprehension, while the light is the necessary attention.[21]

With these ingredients, the student wishing to penetrate the mysteries of science "must proceed from one object to another in accordance with a suitable method. For thus he will apprehend everything surely and easily." The eye, the object, and light were the outstanding words chosen to explain the process.

Not even Bacon, the most lucid of all in outlining his method, gave a practical, working blueprint for experimentation. With his fellow utopists, he stated an exceedingly high goal: "The end of our Foundation is the knowledge of Causes, and secret motions of things; and the enlarging of the bounds of Human Empire, to the effecting of all things possible." Regardless of all-inclusive purpose, he attacked the serious procedural problem with clarity and simplicity. There is no utopia that concentrates more completely on the organization of research than Bacon's, and there is probably no utopia that ever had more influence on men actively engaged in scientific research.

For his inability to limit a project he has been severely criticized, yet specific boundaries might have seriously reduced the impact of his thinking in his own age. As it stood embracing "everything," the promise could not have been greater to others of more scientifically inclined temperaments. His plans seemed sensible enough for the Royal Society to acknowledge him as their inspirational founder and guide. To Robert Hooke, author of the *Micrographia* (1665), who worked with Boyle in perfecting the air pump and examining the process of combustion, Bacon was a great benefactor. In the pursuit of knowledge, Hooke wrote, the intellect

> is continually to be assisted by some Method or Engine, which shall be a Guide to regulate its Actions, so as that it shall not be able to act amiss: Of this Engine, no Man except the incomparable *Verulam*, hath had any Thoughts, and he indeed hath promoted it to a very good pitch.[22]

Many such compliments can be found in his own century, but his position in the longer history of science has been variously evaluated. He has been both condemned and worshiped. Charles Darwin named him as his forefather, went through the collective process himself, and formulated the evolutionary theory. More recently, modern scientists like René Dubos and Loren Eiseley have reexamined Bacon's thought in terms of their fields of biology and anthropology. They, along with Alfred North Whitehead, have emphasized the importance of Bacon's general approach to the problems of nature—which approach has been engulfed by the physical line of thought in which a passive nature is being acted upon externally by forces. Whitehead said:

> I believe Bacon's line of thought to have expressed a more fundamental truth than do the materialistic concepts which were then being shaped as adequate for physics. We are now so used to the materialistic way of looking at things, which has been rooted in our literature by the genius of the seventeenth century, that it is with some difficulty that we understand the

> possibility of another mode of approach to the problems of nature.[23]

So Whitehead, in the interests of relating once more the concrete object to the concept, would support Bacon's approach in preference to the mechanistic line developing in the same century.

Yet, as has often been pointed out, with the exception of Bacon's work on the nature of heat, he did not produce results of consequence in the more strictly physical sciences. Charles Singer comments on Bacon's "experimental ineffectiveness" and the problems inherent in trying to deal with all facts.[24] Choice, which is judgment, must operate, and in this Bacon seems inept, though he realized its importance and stated that the first collection of instances was not to be an indiscriminate amassing; it was to include those instances in which a certain nature is presumed to be present; it was a sorted natural history. Again, his policy of exclusion or negation was a safeguard. If he was unable, for whatever reason, to apply his own method significantly for scientific results, he nevertheless contributed effectively to thought and general approach. He convinced the next generation of the need to clean out past practices, of the difficulties in arriving at truth, the "daughter of time" and not authority, of the importance of removing theological questions from the laboratory without in any way impugning God's omnipotence, and he convinced them sufficiently to join the effort. We are not likely to obtain a consensus of opinion on Bacon's role in subsequent scientific development, but for the utopists of his day, at least, he was a philosophical leader whom they ardently followed, and he set the dominant pattern, even though their scientists' work was frequently not so carefully organized and planned as it was in *New Atlantis*.

The method that Bacon expounded in the *Novum Organum* was described in full operation in Salomon's House, where practical results had already been achieved and more were constantly expected. In assigning personnel and specifying the chronology of their duties, Bacon outlined his system of collection, analysis or experiment, and results. The basic task of collection was carried on by three groups of men: the Merchants of Light who sailed to foreign shores and brought back materials and patterns for experiment; the Depredators who carried on their research in books; and the Mystery-men who examined experiments in the mechanical arts, liberal sciences, and techniques outside the arts. This last group surveyed the work of craftsmen and artisans whose ingenuity had often solved problems through practical necessity when the more formal laboratory routine had failed. Meanwhile, "Pioners" or Miners were trying original experiments that seemed promising to them. The second step in the orderly series was performed by Dowry-men or Benefactors, who evaluated the accumulation of "particulars" from the points of view of practical use and further theoretical knowledge. Under the direction of the Lamps, the next round of experiments was executed by the Inoculators, and the Interpreters of Nature at last elevated the actual discoveries of fact to the high level of axioms or aphorisms, that is to say, truth.

In *New Atlantis* there was for the first time in utopia a *cooperative* system for the advancement of knowledge as well as the first evidence of specialization in the worker's function. This division of labor was a most important idea, because it meant that many hands could do more, do it more quickly, and thoroughly. And the concomitant was directly implied: namely, that with such organized effort, society need not await the chance appearance of an individual genius.[25] Furthermore, Bacon provided no training courses for the specialized worker, which was another unique innovation in his system. He counted heavily on his method to guide the industrious person of common sense to the point of discovery. In drawing a straight line or a circle, much depends on steadiness and practice of the hand, he said, "but if with the aid of rule or compass, little or nothing; so it is exactly with my plan."[26] His procedure was such that all men should be capable of employing it, which immediately leveled the tasks and encouraged any man to believe he could perform successfully and aid the great endeavor.

Only in one area did he advise additional guidance. In view of the heavy dependence on the visual element, it was naturally important to train the human "tool" to guard against deception and errors that mislead the senses. Hence the utopian worker, especially in the systems of Bacon and Joseph Glanvill, received detailed instruction. In Salomon's Foundation, there are "houses of deceits of the senses" where one may see represented

> all manner of feats of juggling, false apparitions, impostures, and illusions; and their fallacies. And surely you will easily believe that we that have so many things truly natural which induce admiration, could in a world of particulars deceive the senses, if we would disguise those things and labour to make them seem more miraculous. But we do hate all impostures and lies: insomuch as we have severely forbidden it to all our fellows, under pain of ignominy and fines, that they do not shew any natural work or thing, adorned or swelling; but only pure as it is, and without all affectation of strangeness.[27]

When Glanvill continued the description of *New Atlantis*, he reemphasized the danger of false impressions, and said that even the utopian was apt to be confounded by his "Complexions, Imaginations, Interests and Affections." So all were cautioned to beware, but, aside from this, the worker was simply to follow the right method.

Of course in the small-scale model of *New Atlantis* as Bacon created it, there were only 36 persons employed in the entire scientific process. Andreae's laboratories, in contrast, seemed to swarm with an unlimited number of workers, so he took care that

> the men are not driven to a work with which they are unfamiliar, like pack-animals to their task, but they have been trained long before in an accurate knowledge of scientific matters, and find their delight in the inner parts of nature.[28]

But then, Christianopolitans did not have as dependable a ladder of investigation as Salomon's House. Neither did the Solarians, the Macarians, nor the citizens of any other new utopia.

Oddly enough, the poet and dramatist, Sir William Davenant, was the only other author to sketch a systematic method. Davenant outlined a process of detailed steps leading from the basic observation of nature to the formulation of truths for Astragon's House in *Gondibert*. But he organized the research process only to the point of assigning categories for the workers. The collectors of specimens of beasts, fish, and fowl were called Intelligencers. They personally selected the area for which they gathered information and made constant observations. These materials and more from distant regions were brought into Great Nature's Office where busy old men, Nature's Registers, recorded the facts regarding each item: weight, measure, habits, motions, and rest. This office prepared data for the learned lord, Astragon, who then employed marvelous powers to reach new truths. The process ended with the mind of Astragon, and society enjoyed great benefits from his remarkable discoveries. Obviously this was no Baconian blueprint, but a diagram of method was suggested.

Usually the utopian scientist's work is simply to observe, "make trial," and record; vital details of the relationship between observation and conclusion are not vouchsafed us. We are simply assured that the method works, and that every day in utopia the barriers of ignorance are reduced. Even the mathematically inclined Andreae describes a labyrinth of laboratories whose projects seem to overlap considerably. In a section on the west side of the city of Christianopolis are located the forge and all industries requiring the use of fire:

> Here in truth you see a testing of nature herself; everything that the earth contains in her bowels is subjected to the laws and instruments of science. . . . If a person does not here listen to the reason and look into the most minute elements of the macrocosm, they think that nothing has been proved. Unless you analyze matter by experiment, unless you improve the deficiencies of knowledge by more capable instruments, you are worthless.[29]

This is the generalized description typical of utopia. And here chemical research, because of the need for fire, goes on, but it also is conducted as well in another laboratory designated for "chemical science." Similarly, in various locations there are workers examining the properties of metals, minerals, vegetables, and animals for the use of the human race. In one place men "learn to regulate fire, make use of the air, value the water and test the earth." Although Andreae was still thinking of matter in terms of the four traditional elements, and he failed to theorize on the composition of matter as Descartes was to do, the Christianopolitans in their wholesale approach to experimentation were taking the first step toward modern scientific analysis: they were asking questions, however vague, examining properties, and trying to test and control them.

The same atmosphere pervaded Bensalem's laboratories and houses for experiment, but functions were more distinctly separated, and instruments and equipment were designed more cleverly for the specific task. Along with the houses experimenting in perspective, sound, perfume, motion, and various other areas, Bacon built high towers, some of them on mountaintops, for meteorological observation and tests for exposure to the sun, and he created low, deep caves for experiments in refrigeration, the conservation of bodies, the study of natural mines, and the production of artificial metals. The highest tower, we are told, and the deepest mine were each three miles at least in measurement!

These facilities reappeared in utopias later in the century. Astragon's House was furnished with similar towers of prodigious height for astronomical research, which was conducted by means of long optic tubes that allowed man to pry into, not merely look at, the moon's face.

> Others through Quarries dig, deeply below
>   Where Desart Rivers, cold, and private run;
> Where Bodies conservation best they know,
>   And Mines long growth, and how their veines begun.[30]

Abraham Cowley, always very conscious of aesthetic values on his utopian campus, also remembered the same Baconian features when he suggested that there should be built

> some place of the Colledge, where it may serve most for Ornament of the whole, a very high Tower for observation of Celestial Bodies, adorned with all sorts of Dyals and such like

> Curiosities; and that there be very deep Vaults under ground for Experiments most proper to such places, which will be undoubtedly very many.[31]

So facilities were structured, often on Bacon's model, and intended to provide for all manner of experiments in an unbelievably wide variety of areas.

Whether in Andreae's laboratories, Bacon's houses, Hartlib's college of experience, or Cowley's chambers, studies were generally conducted on a grand scale. No one was worried about a precise definition of the object or purpose of a particular experiment. Remember, too, the terminology, the jargon, of science had not yet been devised to make explicit a certain detail. One vocabulary sufficed in all fields, so words seeming vague in definition to us now were often, literally, the only expression available to the 17th-century author. Recognizing this fact, we may better evaluate their words, their intentions and plans, and conclude that the scientific "spirit" surrounded their endeavor; the attitude was forming that all is knowable if man will busy himself about it; and the effort required organization, procedure, controlled situations and, above all, materials.

Collections

For many utopians, method was primarily a matter of making enormous collections to be conserved in museums, botanical and zoological gardens. Their emphasis on the collection of facts was most important, as Professor A. R. Hall reminds the student of scientific history:

> Pure fact-collection (the first stage in Bacon's system) has been a most important fraction of all scientific work up to the present time. . . . It is true that the main course of physical science in the seventeenth century ran in a very different direction, that in the new mechanics of Galileo the plodding fact-gathering imagined by Bacon had little significance; elsewhere in science, however, where the organization of ideas was less advanced and the material far more complex and subtle, the straightforward acquisition of accurate information was a more

> fruitful endeavour than premature efforts at conceptualization.[32]

In this interest, utopists had many predecessors beginning as far back as the Alexandrian age with its great library and museum, which differed in its research function from the general libraries that had existed before. Not only did Alexandria start with the nucleus of Aristotle's library and build the greatest collection ever seen at the time, but it housed

> rooms for lectures and study, dissecting rooms, an observatory, a zoo, a botanical garden—in a word all the material requirements for the anatomical, astronomical, biological, botanical and philological studies that were destined to make such rapid progress there.[33]

Europe, with and without knowledge of this tradition it followed, developed similar interests primarily in the 16th century, when the building of great collections flourished again, together with careful observation, realistic detail, and principles of perspective in drawing. Leonardo da Vinci was, of course, the most famous of men in this respect. But all around him others, less well known, were assembling large collections of plants and herbs on which medical practice largely depended, curiosities and rarities brought back by travelers and explorers from the corners of the earth, and animal forms of interest to early "biologists" who could continue to pay their respects to Aristotle, for his biological studies were never subject to the ruthless criticism his cosmology and astronomical theories received. Collections covering the animal, vegetable, and mineral kingdoms were extremely popular among the people whose imaginations were aroused by the unusual forms never before seen. Encyclopedic naturalists, like Leonhard Fuchs, Guillaume Rondelet, and Aldrovando,[34] following in the footsteps of the great Roman compilers, often presented phenomena in printed volumes, exhaustive in detail and drawings and, more important, comparative in approach. These were valuable materials to be studied, and more than once these compendious volumes stimulated a searching question later answered; they were not regarded, as were the Roman collections, as the final answer, but instead, as a summary of progress-to-date.

By the end of the 16th century, Bologna had a great botanical garden, Paris had another as well as a menagerie, and colleges were adding museums of curiosities and building greater libraries. This feverish activity of discovery and compilation thoroughly invaded the new utopia, where the museum was a point of central interest for the visiting tourist. The tour moved through miles of corridors lined with objects while the guide gave a constant description. Collections could, of course, be complete in utopia, and every known object and phenomenon, even hail and lightning, were displayed. Utopian societies arranged their collections in various ways: some were classified and located near the laboratory for use; others remained in general metropolitan museums serving all citizens. Francis Bacon very intelligently installed departmental collections in immediate proximity to the house concerned with them. Thus the assembled objects were related to the special studies in sound, heat, perspective, engines, and perfumes.

Andreae had a somewhat similar arrangement. Outside the gate of Christianopolis stood the drug supply house, which had a more carefully selected collection than any other place in the world. This pharmacy had been made into a veritable miniature of all nature:

> Whatsoever the elements offer, whatever art improves, whatever all creatures furnish, it is all brought to this place. . . . For how can the division of human matters be accomplished more easily than where one observes the most skillful classification, together with the greatest variety.[35]

Lest the reader should think this is the central depository of collections, Andreae hastens to present the elegant hall of physics located nearby:

> For natural history is here seen painted on the walls in detail and with the greatest skill. The phenomena in the sky, views of the earth in different regions, the different races of men, representations of animals, forms of growing things, classes of stones

> and gems are not only on hand and named, but they even teach and make known their natures and qualities. Here you may see the forces of agreement and of opposition; you may see poisons and antidotes; you may see things beneficial and injurious to the several organs of man's body.[36]

Campanella employed the same technique of wall representation for the Solarians' collection of phenomena and, though it embraced the world and all learning, it was not an uncatalogued monstrosity. Categories were limited and logically placed in accordance with nature's hierarchy as it was then conceived. Yet only three of the four basic elements were included in the exhibits. The omission of fire from the display of the elements was immediately noticeable and indicated how closely Campanella followed scientific developments in at least some areas. Fire was the first of the old elements to go. And Campanella had accepted the veracity of Tycho Brahe's work proving that fire, the purest element, was nonexistent in heaven's highest sphere, its traditional home.[37]

While he made such concessions to recent discoveries and explained the qualities and habits of each item, even mounting into the wall the actual object wherever possible, Campanella adhered to his belief in correspondence or resemblance between objects:

> I was astonished when I saw bishop fish, chain fish, nail fish, and star fish exactly resembling such persons and things among us. There were sea urchins and molluscs; and all that is worth knowing about them was marvelously set down in word and picture.[38]

Nevertheless, his organization of materials surpassed his fellow utopists' in its highly centralized structure—and perhaps in his imaginative inclusion of such distinctive items as the only phoenix in captivity—but it can scarcely be said that the other utopists were far behind in fertility of imagination.

Sir William Petty located his *theatrum botanicum* within the walls of his ideal college near the teaching hospital. An apothecary supervised the collection, but his professional interests certainly did not limit the nature of the objects included. There were cages for all sorts of strange beasts and birds, ponds for exotic fishes, museums for rarities, "natural and artificial pieces of antiquity, models of all great and noble engines . . . the most artificial fountains and water-works . . . an astronomical observatory," together with the rarest paintings and the fairest globes. Petty explained that "so far as is possible, we would have this place to be the epitome or abstract of the whole world." No utopia was without a collection or museum, whether it took the form of a "physick garden," a "repository of nature," or Nature's Nursery in Astragon's House, which contained more specimens than the Garden of Eden because man's art had created innumerable new plants. Here was no return to past perfection; the utopia was much better and far advanced in comparison.

In the ideal society, the collection provided the materials for the laboratory table and the classroom. The utopian worker or student examined, tested, and observed in order to imitate nature, and skillfully proceeded to re-create her effects. The scientist was the imitator *par excellence*. In Andreae's chemical science laboratory, for example, "the ape of nature has wherewith it may play, while it emulates her principles and so by the traces of the large mechanism forms another, minute and most exquisite." The implication is that once the worker, the "ape of nature," has mastered her principles, he will control his own environment and live more easily and pleasantly. It is Bacon's point that man will command nature only by obeying her.

Besides providing objects for examination, utopian collections constituted a record of man's accomplishments. In the early part of the 17th century, there was a real need for such records, often expressed in the desire for natural histories encompassing everything known. And, indeed, in a world in which communications were poor and no scientific journals existed, when newspapers were occasional fly sheets, when academies had not yet established a systematic interchange of opinions and reports, the need for compilations of knowledge was naturally felt. It was not only to consolidate the past, a security measure, but to illuminate the present and future as well, showing the next need, the area for questioning and experiment.

Diverse plans were proposed to make knowledge more generally available and to assemble facts from distant lands. The Merchants of Light were sent out from New Atlantis to collect scientific data abroad. Bacon has often been thought to be the creator of such emissaries to foreign countries. He was not, however, the originator of this plan, nor was he the last to suggest it. Plato had advocated the selection of agents from among wiser and older citizens, assigning them the task of learning the laws, education, and customs of other lands so that their own country might be improved.[39] Robert Burton in his "poeticall commonwealth" employed foreign agents for Plato's purpose and cited the *Laws* as his source:

> I will have certaine shippes sent out for new discoveries every yeare, and certaine discreet men appointed to travell into all neighbour kingdomes by land, which shall observe what artificiall inventions, good lawes are in other countries, customes, alterations or aught else, concerning warre or peace, which may tend to the common good.[40]

Ambassadors were also sent out from the City of the Sun by Campanella with the broad purpose of discovering the good and bad of other nations of the earth. From their representatives, Solarians had learned interestingly that "explosives and printing were known in China before they became known among us." And in Abraham Cowley's *Proposition for the Advancement of Experimental Philosophy*, four "Professors Itinerant" were assigned to Asia, Africa, Europe, and America:

> There to reside three years at least, and to give a constant account of all things that belong to the Learning, and especially Natural Experimental Philosophy of those parts.
>
> That the expence of all Dispatches, and all Books, Simples, Animals, Stones, Metals, Minerals, &. and all curiosities whatsoever, natural or artificial, sent by them to the Colledge, shall be defrayed out of the Treasury. . . .
>
> That at their going abroad they shall take a solemn Oath never to write any thing to the Colledge, but what after very

> diligent Examination, they shall fully believe to be true, and to confess and recant it as soon as they find themselves in an Errour.[41]

What Plato had advocated as a means for generally surveying other countries developed into a specialized assignment to collect knowledge regarding the new science.

Gerrard Winstanley thought that the exchange of information was essential within the borders of the utopian land itself. He established an office of postmaster[42] so that one part of the country might learn of the work going on elsewhere. He explained the function:

> Or if any through industry or ripeness of understanding have found out any secret in Nature, or new invention in any Art or Trade, or in the Tillage of the Earth, or such like . . . when other parts of the Land hear of it, many thereby will be encouraged to employ their Reason and industry to do the like.[43]

Concurrent with the exchange system at home and the collection of data from abroad were grandiose local programs for amassing facts. Although this was recognized to be a huge job and occasionally realists cautioned against it, as when friends advised Hartlib and Comenius that times were not ripe, that interests were diverted elsewhere, that they were a bit naïve and credulous in their efforts to collect all knowledge, it was still undertaken in the optimistic belief that its chances for success within a generation or two at most were excellent. Comenius replied to the critics and admitted that "even now men will say that we are drunken and dreaming," but he added, "the answer is to make all men drunk."[44]

Samuel Hartlib, whose brain teemed with designs for compilation, presented a popular plan for an *Office of Publick Address.*[45] This was another clearing house for many types of information with the important function of serving as a "Register of Ingenuities, and Matters commendable for Wit, Worth and Rarity." Records were kept of new feats "in physick, mathematicks, or mechanicks; or a method of delivering sciences or languages." The communications office housed oddly assorted collections—roots, instruments of mathematics and astronomy, anatomies of creatures dead or alive, and rare goldsmiths' work. Hartlib concluded his outline with the draft of an act for Parliament, just as he dedicated his utopian *Macaria* to that august body, hoping for action. And he spent a great deal of time searching for a suitable location for the proposed office. Two years after the publication of his plan, he wrote his friend Robert Boyle that he had been unable to obtain Vauxhall, which he considered an ideal place for the office. Obviously, he was serious about it.

This was the same office that William Petty intended to support from his invention of the tool for "double writing," and he started his *Advice* with a strong recommendation for the "institution of an office of common address, according to the projection of Mr. Hartlib." The essential needs of the new science would be met in such an office,

> where men may know what is already done in the business of learning, what is at present in doing, and what is intended to be done: to the end that, by such a general communication of designs, and mutual assistance, the wits and endeavours of the world may no longer be as so many scattered coals, or firebrands, which for want of union are soon quenched; whereas, being but laid together, they would have yielded a comfortable light and heat.[46]

As matters stood, Petty thought some men were puzzling themselves to reinvent what had already been invented, and others were confronted with difficulties that another scientist could assist in solving.

One of Petty's plans was for a great history of trades, which would reveal all "practised ways of getting subsistence," and which might aptly bear the title, "The Golden Fleece; or a great Description of Profitable Callings."[47] This gigantic assignment was given to one man, but Petty recognized that the editor would have to start the compilation in his youth, and then pray God to give him life long enough to finish it. The editor was charged to beware of false information and to observe directly the facts for inclusion. Double-checking by experts in each area was necessary because the history was to be more than a reference book: it was the basis for further "philosophations" and inventions. An apprentice studying this history could reduce his period of training from seven years to three or four. The material included was to be sufficiently descriptive so that a student would not spend years of preparation only to learn he was unfitted by nature or ability for the job. Moreover, the remarkable history was to serve as a laboratory manual giving reasons for the use of each instrument and each step in the scientific process. This was extremely valuable if men were to know what they were doing and perform more effectively. It has been pointed out that such "Histories of Nature, Arts or Works" as those sponsored by the Royal Society provided for the first time scientific descriptions of craft technologies that had often been kept mysterious in the past to protect the craftsman's art.[48]

Petty visualized the history of trades as a volume dealing with "nature vexed" and artificial, to be followed by a second volume based on Bacon's model to treat "nature free." Although he considered himself Bacon's disciple and worked to achieve the Lord Chancellor's dream, he added detail and concentration to his master's plan. While he embraced tremendous areas of knowledge, the boundaries were limited, and his thoughts were certainly more solidly grounded than those of his immediate predecessors in the utopian fields, perhaps because he was acutely conscious of cost.

The great utopian compilations of facts and objects—whether assembled in museums or presented as Bacon's or Petty's encyclopedic natural history—were to be an epitome of the whole world. Facts were gathered from all foreign lands by Bacon's Merchants of Light, Burton's foreign agents, Campanella's ambassadors, and Abraham Cowley's "Professors Itinerant." Meanwhile, utopian citizens at home were tirelessly collecting and studying. No object was too small or too great for inclusion in the all-important collection. These materials were valuable not only as a current record of knowledge, indicating areas for further exploration, but, more significant, they were basic materials for the classroom, laboratory, and workshop. Here the process started.

Then, building ever upward from the concrete object, the utopian worker was to reach the universal truth. There was no question of adopting the deductive approach; that was generally repudiated. And, with the exception of Andreae, who remembered the powers of quantitative analysis, the utopists generally ignored the great possibilities of the mathematical method, paying only lip-service to the field. For them, the right method was Baconian and heavily dependent on techniques of observation, analysis, and experiment. The visual sense was of key importance, and proof had to be apparent to the senses. This was Glanvill's final "report" to the human faculties that could be counted upon to reveal impartial truth after free examination and trial with proper precautions. Thus, in these early days of modern science, the utopists emerged as champions of the empirical method, the most promising route to the splendid future they envisaged.

When method or process became a *consuming* interest for the utopist, the ideal society was no longer static in perfection, no longer a Dantesque paradise or a projection of the final complete ideal. It was well beyond the ancients' attitude toward change and possibility; it was now a creative process with man's mind in control and belief in his heart. In the 17th century, utopians optimistically adopted the idea of progress; consequently, their society became dynamic. Because they were actively searching for new knowledge and working toward the conquest of nature, they could not conceivably believe that their society already held the final answers to man's happiness. It was promise that they underwrote. The answers were still to come in the future. The utopists' main preoccupation was with the method by which the future might be forced to disclose them.

VIII

INVENTIONS AND RESULTS

Utilitarianism joined forces with the empirical method to create the mighty design that was to rule more than utopia. Even as matter, the object, or material itself was basic in the experimental laboratory, so the material side of existence gained increasingly in importance. Throughout their stories of the way in which the new world was to be brought about, the utopists described enticingly the benefits that encouraged workers to apply more energy to the process. The utopians living in the ideal communities, of course, were unusually fortunate in being able to display many concrete results before the process they used to build with was completed. They did not have to wait for the materialized product; nor were they compelled to say generously that at least their children would see the result, though they anticipated for them an even more remarkable world than their own. When utopists took the imaginative flight to the promised land, they guided their course by the principle of utility. Bacon went so far as to say that "what in operation is most useful, that knowledge is most true." The main gauge to the value of a discovery was whether man's life became easier, healthier, and more pleasant. Thus, the 17th-century utopia introduced the age of applied science and technology.

Today we tend to think of technology as an outgrowth of scientific knowledge, but historically the reverse is true: technology, the practical solution to a problem, often preceded scientific theory and provided the working example from which the principle was later formulated. Galileo intelligently explained the relationship at the beginning of his famous *Dialogues Concerning Two New Sciences*. Salviati, the chief character in the dialogues, speaks:

> The constant activity which you Venetians display in your famous arsenal suggests to the studious mind a large field for investigation, especially that part of the work which involves mechanics; for in this department all types of instruments and machines are constantly being constructed by many artisans, among whom there must be some who, partly by inherited experience and partly by their own observations, have become highly expert and clever in explanation.

And Sagredo, agreeing with the idea, answers him:

> You are quite right. Indeed, I myself, being curious by nature, frequently visit this place for the mere pleasure of observing the work. . . . Conference with them has often helped me in the investigation of certain effects including not only those which are striking, but also those which are recondite and almost incredible.[1]

The "causal connection" making the phenomena possible was the vital point necessary to understand if man were to assume the controls, and this was what earlier inventors had frequently failed to examine. Hence, though technology may be said to have existed since man began to solve practical problems, it was not until the 17th century that formulations of law were stated to explain the operational facts.

Another reason for technology's precedence in development was the powerful economic factor: support was available for a Galileo if he would concentrate his efforts on practical matters such as Venice's arsenal and cannon foundries. Again in dialogue, the spirit of this issue was amusingly caught, this time by the modern playwright Bertolt Brecht in his play *The Life of Galileo*. In an early scene, the curator is dickering with the great scientist over financing his research:

> *The Curator.* And as far as the material side is concerned, why don't you invent something else as pretty as your wonderful proportional compasses which—*he counts off on his fingers*—enable one without any mathematical knowledge to protract lines, calculate compound interest on capital, reproduce ground plans in varying scales, and determine the weight of cannon-balls.[2]

Admittedly the suggestion is stupid, and Galileo retorts: "A toy!" Other tales bear out the feeling that such ridiculous requests were fairly typical of those not understanding the new thought, but devoted to practical gadgets. Yet in truth Galileo's work often rested upon the solution of a realistic need. The same was true of the founder of mechanics, Archimedes in the 3rd century B.C., who now was reinstated with worship for his remarkable discoveries.

Failures in technology have more than once shown the need for further theoretical work just as success in application has confirmed a principle. Interaction between the two types of research, since the 17th century, has been reciprocal in scientific progress, and this the utopists recognized in their own day. While the focus was on applied science, they did not preclude a concern for purely theoretical research. Utopians were not so lacking in foresight.

Francis Bacon's Bensalemites have been censured for this, but they deserve acquittal. Although emphasis was on utilitarian benefits or "fruits" of experimentation, Bacon did not overlook experiments of "light," which led to general truths. To be sure, at the second rung of the ladder of investigation, the Dowry-men or Benefactors applied what had been learned to date. They "cast about how to draw out of them [the collected facts] things of use and practice for man's life." But, though "fruits" were drawn from experiments at this early stage in the method, the process itself was continued on to generalized conclusions—axioms. For Bacon, progress in power and progress in knowledge were two aspects of the same thing.[3] The other utopists, like Bacon, made no conscious choice between applied and theoretical research, assuming both to be in the same forward movement. Yet they were very eager to get results and so concentrated on science applied for man's betterment. Their desire to increase knowledge for its own sake was seldom mentioned, no doubt because they had had enough of that in the abstractions of medieval philosophers and disputatious schoolmen. Knowledge now had to materialize in evidence—in improvement of man's daily life and in his control of the forces about him.

Utopians were especially encouraged to exhibit their inventive skills because prowess in discovery had come to indicate progress itself. Inventions were yardsticks that measured the ancients' knowledge and the moderns' progress. When Campanella claimed for Solarians more "history" in the last century than in the world's 4000 years before, he gave as evidence those "stupendous inventions"—the compass, the printing press, and the harquebus. And Bacon, considering the importance and consequence of discoveries, stated the claim even more strongly:

> These are to be seen nowhere more conspicuously than in those three which were unknown to the ancients: . . . namely, printing, gunpowder, and the magnet. For these three have changed the whole face and state of things throughout the world; the first in literature, the second in warfare, the third in navigation; whence have followed innumerable changes; insomuch that no empire, no sect, no star seems to have exerted greater power and influence in human affairs than these mechanical discoveries.[4]

Scientific Instruments

These three signs of progress were constantly cited by writers throughout the century, but by the time Joseph Glanvill evaluated "Modern Improvements of Useful Knowledge,"[5] the list had been expanded to include five new and important instruments: the telescope, microscope, thermometer, barometer, and the air pump or, as Glanvill called it, "Mr. Boyle's Pneumatick Engine," which conclusively disproved the fancy of a *fuga vacui*.[6] Glanvill might very well have added to his list the pendulum clock, which enabled man to measure small-time intervals.

The utopists were well aware of the value of such instruments to aid the senses by extending their powers of observation and by protecting them from sensory deception. In societies basing their scientific work with such total confidence in the senses, these instruments were naturally of immense importance. No one spoke more clearly or persuasively than Francis Bacon on the need of "instruments and helps," though he habitually cautioned, as we would expect, against depending on them too much. The Father of Salomon's Foundation explained at length the program of sensory research and the instruments used in the experimental "houses." Work was being done on optical problems that had long interested men in the real world and had received new impetus from the "optick tube" made popular by Galileo. Remarkable results are already visible. The workers can make far-off things near and, in reverse, represent "things near as afar off." They have developed spectacles or glasses "far above" those being used elsewhere, and most amazing:

> we have also glasses and means to see small and minute bodies perfectly and distinctly; as the shapes and colours of small flies and worms, grains and flaws in gems, which cannot otherwise be seen; observations in urine and blood, not otherwise to be seen.[7]

This valuable instrument remained unnamed in the *New Atlantis*, but at least the workers were familiar with its abilities. Bacon probably did not know the term, microscope, which was apparently first used in 1625 by Giovanni Faber in a letter concerning Galileo's microscope.[8] The idea of a magnifying glass had been current for centuries, however, and Bacon may simply have enlarged on its possible values to scientific research. In any case, he certainly deserves to be called a "spiritual ancestor of microscopy in England," as Marjorie Nicolson has described him. It was not until 1660, or some 36 years after Bacon wrote the *New Atlantis*, that the microscope was recognized as an important aid to learning in botany, zoology, microbiology, and other such subjects. And it was in this year that it reappeared in R. H.'s *Continuation* as that "rare *Microscope*, wherein the eyes, legs, mouth, hair, and eggs of a Cheesmite, as well as the bloud running in the veins of a Lowce, was easily to be discerned." Though utopists in the intervening years had not mentioned the microscope, by 1660 it was no longer unusual to encounter it or the selenoscope, as they termed the telescope, which was used in the study of selenology or the science and movements of the moon.[9]

The telescope, of course, made frequent appearances in

17th-century utopias, since it was already in use and had more than excited men by its demonstrated powers. Andreae mentioned it, and Campanella's citizens who were living some eight years before Galileo's optic tube, anticipated the discovery of "a glass by which to see the hidden stars." Prevenient as it seems, there is evidence that this passage was added to the early manuscript after Galileo's announcement.[10] Davenant directly reported that the study of optics at Astragon's House had progressed wonderfully and that "Moderns are become so skill'd they dream of seeing to the Maker's throne." In his *Advice,* Sir William Petty suggested that youth be taught to grind lenses and that men study "pure mathematicks, the anatomy of the eye, and some physical principles, concerning the nature of light and vision, with some experiments of convex and concave glasses."

There was some basis for the adoption and perfecting of work in optics, but utopians had much less reasonable grounds for many of their unique inventions. Ingenious aids to probe the world of sound and transmit the voice across great distances were devised. Here Bacon[11] again projected many instruments well ahead of scientific actuality. While Campanella's society was expecting mechanical devices through which they could hear the music of the spheres, Salomon's House had perfected hearing aids for the deaf. Governmental funds in Bensalem also supported the development of instruments that made small sounds "great and deep" and "artificial echoes, reflecting the voice many times, and as it were tossing it." Perhaps Bacon foresaw the microphone and modern acoustic inventions as well as the telephone in the means "to convey sounds in trunks and pipes, in strange lines and distances." But Bacon was not singular in this. "Sympathetic" devices for communication were exceedingly popular in the century and many were proposed and believed feasible. Bishop Francis Godwin tells of a "brazen pipe" put into a wall connecting towers and a castle that he remembered from the historian William Camden's story of this practice in northern England by the inhabitants of the Picts' Wall. Godwin describes it in his *Nuncius Inanimatus or the Mysterious*

Messenger in Utopia (1629), which briefly summarizes the efforts of the human race to communicate and asserts that by his new device—the "inanimate nuncio"—man may quickly transfer the voice great distances. Realistically, Godwin admits that the device will cost much and, of course, require considerable preparation and design "for perpetual use," but no magic or unlawful acts will be needed. Sound with definite meanings assigned to it (the principle of the Morse Code) will be used, and he begs the reader: "Do not rashly pronounce this proposal impossible."[12]

R. H. was more explicit in details for his "most admirable" invention that was to carry men's voices for many miles:

> Two needles of equal size being touched together at the same time with this Stone, and severally set on two tables with the Alphabet written circularly about them; two friends, thus prepared and agreeing on the time, may correspond at never so great a distance. For by turning the needle in one Alphabet, the other in the distant table will by a secret Sympathy turne it self after the like manner.[13]

It is "sympathy"—not modern sound waves or radio—on which promises are predicated. Seldom are we given sufficient detail to evaluate utopian ingenuity in terms of actual modern developments. Yet their imaginations described, and they firmly believed it would be done. It is important to realize what, in this case, is the stuff that "dreams are made on," and that these actors were not spirits that melted into thin air. Their utopian imaginations were well prepared for the results that have come, perhaps more prepared than we are today for visits to outer space.

Other utopists extolled resourcefulness and mechanical aids in even looser, more sweeping terms. Campanella's Solarians simply allocated the innermost wall of their city—surrounding the temple in the center—to an exhibit of instruments and mechanical arts. Christianopolitans set up a special laboratory for mathematical instruments, which appeared again in Cowley's ideal college. But aside from these few instances in which mechanical or mathematical aids were given a special physical location, "instruments" were seldom differentiated from "inventions" and "discoveries." The three terms were used interchangeably in the 17th century. Columbus and Magellan were "inventors" and took their positions in galleries honoring men who had contributed to new thought. "Discoveries" could be an improved fruit or a new method of preserving, and the much-heralded telescope was called an invention, a discovery, and an instrument. Our present-day distinction was unknown. Now invention in the sense of discovery is considered archaic and obsolete. We generally call new technological developments "inventions" while the term "discovery" is reserved for greater truths such as Newton's law of gravitation. But for utopists of the century that formulated many of our modern notions, the terms were equated and expressed the belief that progress in power and progress in knowledge were inseparable.

The goal was wide enough to embrace any "useful kind of knowledge." Some astonishing "gadget" inventions were introduced, but most innovations had far-reaching effects. Although imaginary inventions poured forth as from a cornucopia, they were never unrelated to the actual problems of society. It was not a helter-skelter pile of bright ideas; instead, the vision was one of plenitude with order imposed by problems of the real world. The utopist exercised his right to correct the ills of humanity, prescribing science for their amelioration.

Agriculture

Major foci for inventions and discoveries were four: agriculture, medicine, industry and mining, communications and commerce. Military technique was also mentioned and occasionally with emphasis as by Campanella, but generally it was a lesser concern because utopians strongly desired peace; so they only referred to a utopian preparedness and a protective location. When considering self-defense or a threat to strategically located neighbors, they employed, of course, the latest methods, and often had secrets withheld from the reader as well as the enemy. In selecting the four general fields for the application of scientific findings, utopians definitely paralleled problem areas that were most acute in England at the time.[14]

The utopist's own experience had been in a predominantly agricultural society, yet in this period of expanding industry and trade, he saw people rapidly moving into urban centers. Sir William Petty estimated that London doubled its population between 1636 and 1676, whereas others claimed that it trebled.[15] One of Samuel Hartlib's five Under-Councils in *Macaria* was to establish new plantations to control surplus population. The increased population together with the shift to cities intensified the need for improving agricultural productivity and marketing procedures. Utopists solved both problems speedily. Though some of their ideas were not introduced for many years, they anticipated many 20th-century farming procedures and practices in animal husbandry and horticulture.

Formulas for producing the improved effects were not given, but complete success was taken for granted. Before charging charlatanism, however, it must be noted that governmental patents were actually granted on descriptions no more detailed than those the utopists presented. David Ramsey and Thomas Wildgosse received a patent in 1618 for "engines or instruments, and other profitable inventions, wayes, and meanes . . . to ploughe grounds without horse or oxen."[16] Again in 1627, William Brouncker, John Aprice, and William Parham registered their claim for what seems to be a similar device, at least in its purpose:

> A most readye and easy way for the earing [sic], ploughing, and tilling of land of what kinde soever without the use or helpe either of oxen or horses, by the labour or helpe of twoe men onely to goe with everie plough that shalbe used, with an engyne or gynn [sic] for that purpose.[17]

If these two contraptions merited patents with no more explanation than this, R. H. should have entered a counterclaim because the Bensalem of 1660 used such a device and its inventor —not one of the patentees—was properly honored. A sculptor had made the "head only of the ingenious *Boniger* erect upon a brazen winged colomn," and an inscription explains: "This is the man that contrived the Horizontal sailes, by which three ploughs may go together, and at one time both plough, sow and harrow."[18] Solarians still plough by hand, but they have a similar invention to lighten their work. In their early society pictured in 1602 they use wagons driven by sail to carry the harvest and "when there is no wind, one beast is enough to draw even a large one—a wonderful thing!" Apparently Campanella thought it over and by the 1623 description of Solarians, their wagons fitted with sails are "borne along by the wind even when it is contrary, by the marvellous contrivance of wheels within wheels."[19] Perhaps these fantastic vehicles operated on different principles (whatever they were) and so all were worthy of patent-status!

No less miraculous than the sailing wagon were the suggestions for increasing soil fertility and general productivity. In this research area, Bensalem showed the greatest advancement during the years between Bacon's report and R. H.'s additions. The island originally had soil of rare fertility and varieties of ground for planting "divers" trees, herbs, berries, and vines. Constant experimentation in grafting, inoculation, and cross-breeding had improved species in the vegetable and animal worlds, and Bensalemites had skillfully speeded up the processes of nature. The Father of Salomon's House tells the visitor:

> We make (by art) in the same orchards and gardens, trees and flowers to come earlier or later than their seasons; and to come up and bear more speedily than by their natural course they do. We make them also by art greater much than their nature; and their fruit greater and sweeter and of differing taste, smell, colour and figure, from their nature. . . .
>
> We have also means to make divers plants rise by mixtures of earths without seeds; and likewise to make divers new plants, differing from the vulgar; and to make one tree or plant turn into another.[20]

Bensalemites know modern horticultural methods and have discovered the generative principle that some biochemists today believe can be learned from conditions on and about the moon, if exploratory rockets do not destroy the chances of research. Bacon's words, however, merely reflected an old belief, one he shared with Telesio: namely, that spontaneous generation was a natural phenomenon in an animistic universe in which plants simply had a coarser spirit than higher forms of life.

His first statement concerning plants that come up out of season and bear more speedily was a valuable solution to a severe problem in his own day and related to his other attempts at refrigeration. An early cookbook published in 1609 gave many recipes for keeping fruit "long fresh" or "drie all the yeare," and among ways to preserve meat, the author had a "most singular and necessarie secret for all our English Navie." This was how to carry meats at sea without that "strong and violent impression of Salt which is usually purchased by long and extreame powdering."[21] The difficulty of preserving food was nearly as serious for those on shore; so repeatedly utopists corrected the situation.

Still Bacon's ideal people had no large-scale governmental programs for agricultural development; they merely featured an experimental station. This R. H. rectified by adding a college of agriculture to each of the three major universities in the largest cities. The most famous of these covered about 1000 acres, and research projects were defined in more careful detail. Soil fertility was now a problem, for the original perfection had apparently decreased through constant use. The Jewish guide Joabin explains:

> We dayly try several experiments of . . . meliorating the Earth with several Composts; as the dry with Marle, the lean and hungry with dung of Pigeons, Man's or Horses, Soot, Seasand or Owse, Chalk, & the Sandy with Mud, the Cold with Ashes; the rich with Brakes, Straw, Seaweeds, Folding of Sheep, &, all which, as we find the Ground, we use and apply to it.[22]

This is "but the quickening of Nature by art," and Joabin states the purpose: "All our study here is to improve a little ground well with little pains and charges. For we conceive the well improving of a small Island better than the conquering of a new large Kingdom." R. H. had his own England in mind for this proposal. His farmers were organized in a centrally planned program and given instruction. Every farm of a certain value had to maintain beehives and a certain number of silkworms, with the proportionate number of mulberry trees. Some trees and bushes were planted for berry yield, others for firewood and fast growth.

For each tree cut down, ten were to be planted in its place—the principle of reforestation has been introduced. Here too a solution has been offered for a serious problem—the depletion of timber resources at the very time when England was building a great navy and establishing herself across the seas. Commissioners of the Royal Navy appealed to the Royal Society for suggestions on "the improvement and planting of timber,"[23] and Thomas Sprat in his *History* said that one of the chief activities of the Society was the propagating of trees.[24] Robert Hooke was ordered by the Society to determine the most "stout scantling" by testing the same wood of several ages, grown in different places and cut at different seasons of the year.[25] Many men were involved in the project and John Evelyn, four years after R. H.'s utopia, printed his *Sylva, or a Discourse of Forest-Trees, and the Propagation of Timber*, which included the concept of reforestation.

It will be noticed that utopists did not differentiate between the farming of plants, flowers, vegetables, herbs, or trees. All were growing in profusion, but under controlled programs. There were fast-growing grapes planted on the south side of farm buildings and in sufficient quantity in R. H.'s Bensalem; so there was no lack of good wines. In growing field crops, farmers rotated the species of grain each year, and limed the ground and corn, mainly to combat mildew and smuttiness of corn and to keep birds and worms away. (Campanella did not use fertilizers in his society because he feared they would rot the seeds and shorten the life of plants just as "women who owe their beauty to cosmetics rather than exercise bear sickly children.") Moss, suckers, mistletoe, and other parasites were carefully pruned from trees in orchards. As a result of such advanced practices, Bensalemites had produced over 200 kinds of apples, and many exotic fruits and berries. The anonymous R. H. may well have been a horticulturist or herbalist judging from the nomenclature and his obvious pleasure in listing the plants growing in physic gardens and, indeed, all over the countryside. In every full-scale utopia, agricultural resources were significantly increased through applied science and experimentation, though no other quite matched R. H.'s *New Atlantis*. Still, by whatever means, each utopia was like Hartlib's *Macaria* in which the "Whole kingdom is become like a fruitful garden."

Medicine

Medicine and health were closely related to the provision for improved food products. Although utopians throughout history had always been extraordinarily healthy, those of the 17th century explained at considerable length just how this wholesome state could be achieved. Eugenics and preventive medicine in diet, exercise, and temperance were not the only means, although these methods, which had solved the health problem for older utopias, were often retained. More important in the new utopia was experimentation in chemistry, anatomy, and pharmacy.

In the *City of the Sun*, citizens clung to traditional customs longer than did other utopians, but dietary habits and exercise prescribed by doctors had not been able to eradicate all disease —only those such as gout, catarrh, sciatica, and colic, believed to come from indigestion and flatulency. Also, they never had asthma, which arises from a heaviness of humor. "Cases of inflammation and dry spasm are more likely among them, and these they cure by bathing and eating wholesome foods." Bloodletting and sweating were used for dispelling "infectious vapors that corrupt the blood and marrow." Solarians seemed to suffer most from fevers and epilepsy, which they called the "sacred sickness" after the manner of the Hippocratic writers on medicine who, incidentally, decided it was not a sacred disorder but rather a natural one whose cause had not been located. Still, for Campanella it was at least poetically sacred and the Grand Master pointed out in the dialogue that epilepsy was a "sign of great cleverness," because from it Hercules, Scotus, Socrates, Callimachus, and Mahomet all suffered. This disorder was cured by prayers to heaven, "fragrances, head comforters, sour things, gaiety, and fatty broths sprinkled with flour." An eclectic treatment, to say the least, and one the patient must have enjoyed for its variety. Preventive practices as well as cures operated on the old theory of four humors and the balance of contraries: "In summer or when they are tired, they spice their drink with crushed garlic and vinegar, with wild thyme, mint, and basil." And, in a later edition, it is added that Solarians ate grapes in autumn, since they were given by God to remove melancholy and sadness.[26]

The cures may have been a bit eccentric, but they were effective, for Solarians generally lived 100 years and a rare few reached 200. This was primarily due to a "secret, marvelous art by which they can renew their bodies painlessly every seven years." Perhaps they had discovered how the blameless Ethiopians managed to last so long, or maybe they were first to find the Bensalemites' Waters of Paradise, which were conducive to health and prolongation of life.

In Bensalem, however, medical treatment consisted of more than health-giving baths or Galenist herbal prescriptions. Dissection of animals had shed light on what surgery could do for man, and doctors had learned how to continue life in the patient "though divers parts, which you account vital, be perished and taken forth." Shops of medicines or "dispensatories" were extensively stocked with "simples" and chemically prepared drugs:

> And for their preparations, we have not only all manner of exquisite distillations and separations, and especially by gentle heats and percolations through divers strainers, yea and substances; but also exact forms of composition, whereby they incorporate almost, as they were natural simples.[27]

Probably Christianopolitans were as well informed on those matters as the Bensalemites, for they based medical study specifically on "physics, chemistry, anatomy and pharmacy." Andreae states that they had all the various instruments necessary to repair the body's defects. Moreover, his society was unique in providing for mentally ill citizens and in the benign treatment recommended for their care:

> Persons whose minds are unbalanced or injured they suffer to remain among them, if this is advisable; otherwise, they are kindly cared for elsewhere. This is what is done in case of the violent; for reason commands that human society should be more gently disposed toward those who have been less kindly treated by nature.[28]

Not until Sir William Petty's *Advice* in 1648 did utopian schemes make another valuable contribution to modern medical practices. Petty's *nosocomium academicum,* "an hospital to cure the infirmities both of physician and patient," anticipated today's teaching hospital. A physician supervised surgeons and their assistants, the nursing staff of "honest, careful, antient widows," the apothecary, and laboratories; and a separate business office was maintained for purchasing, bargaining with the workmen, and the keeping of accounts accurately for presentation to the board of curators.

The elaborate details of the patient's case history would surely delight the modern physician. Students accompanied the doctor on his rounds twice a day and made entries in the record. Nurses were instructed to "observe all remarkable accidents happening in the night; as whether they [patients] raved or talked much in their sleep, snored, coughed, etc. all which they shall punctually report to the physician." Facts were taken not only from the ward examination, but also from analyses in the laboratory and "anatomical chamber." For each treatment administered, results were meticulously entered in the record. Petty was modern in his recommendations, yet in a sense he was only bringing back to the light what once had been tried in the early days of Greek medicine. As Campanella was in the Galenist tradition, so Petty was reviving the Hippocratic approach whether he knew it or not. While Hippocratic doctors depended mainly on surgery, diet, and exercise, their writings include an important collection of case histories, which Benjamin Farrington has appraised:

> These record the clinical observations of the physician throughout the whole course of various illnesses, some lasting

> several weeks. The different symptoms and phases of the diseases are recorded with business-like accuracy; the treatment, which generally amounted to no more than keeping the patient in bed and feeding him on slops, is set forth; and the result, which in the majority of cases was death, faithfully reported.[29]

It may be hoped that more of Petty's patients survived and enjoyed more appetizing meals. Petty no doubt studied the classical volume during his medical training, but he was also applying his statistical method to medical matters. He knew full well the contemporary problem of quack doctors with inadequate training; so he would educate the physician as well as give better care to the patient.

Certainly it was this problem that aroused R. H., who, for the first time in utopia, required a physician's prescription before drugs could be administered even in one's own family. He was most distressed at

> the death of many by the errors of a few unskilful Empericks, who not rightly understanding the true aeconomy and state of their patients bodies, or finding out the peccant humours and parts worst affected, commonly expell humours less offensive to their final prejudice. The like care is taken in all Cities and Towns that no *Apothecaries*, *Chirurgians*, *Women* or *Empericks* shall administer Physick to any patient or prepare it, not so much as to their own husbands, wives, Children or Servants, without the Physicians Special advice and direction appointed for that place.[30]

Such a requirement in Bensalem coming at this time (1660) directly corroborates Robert Merton's evidence for the physician's rising prestige and the widespread concern over medicine and surgery at midcentury.[31] Outstanding men like Thomas Sydenham and Richard Wiseman had elevated surgery to new heights; and the fact that many famous scientists—Copernicus, Harvey, Gilbert, Brouncker, and Lister—were also doctors had encouraged the upswing in public opinion for the field of medicine. Hence, this was a period of aggrandizement for physicians. The earlier utopists had suggested this change in attitude to come. Campanella had placed a Medic immediately under Wisdom who had charge of the "sciences and of all the doctors and masters of liberal and mechanical arts." This was a most respected caste, and the doctor vitally controlled many aspects of the environment for the people. Bacon and Andreae had placed high the practice of medicine, and Hartlib had put physicians in charge of his College of Experience, which was the heart of scientific research in his kingdom. So the dreamers had written with eyes wide open for the advancement of medical arts without which utopia could not last long.

Industry and Mining

In both agriculture and medicine, utopists had extended the scientific attitude, and transferred the new method of fact-collections, observation, and trial from the laboratory to the corn field and the hospital. In the other two areas singled out for special attention—industry and communications—the projected improvements were even closer to modern notions of technology, that is, science applied to the development of specific inventions. It is in the various workshops and laboratories that we sense the present-day atmosphere of industrial research and the production of new materials.

Half of the entire island of Caphar Salama was reserved for workshops; and inside the city, as we have seen, were specific zones designated for the various industries. This plan has been regarded as a forerunner of modern city-planning, but it is also a remnant of the old guild system in which crafts were located in certain districts of the medieval town.[32] What went on inside the utopian factory was more unusual:

> For here on the one side are seven workshops fitted out for heating, hammering, melting, and molding metals; while on the other side are seven others assigned to the buildings of those workmen who make salt, glass, brick, earthenware, and to all industries which require constant fire.[33]

Raw materials were placed near the appropriate factory. "Here," as Andreae says, "is practical science."

Winstanley's society was analogous to Caphar Salama in maintaining this close connection between the crude substance and the finished product. The second of the five "Fountains" of learning was "mineral employment" which was the search for mines of gold, silver, and other metals of the earth. "Here all Chymists, Gunpowder makers, Masons, Smiths, and such like, as would finde out the strength and power of the Earth, may learn how to order these for the use and profit of Mankinde." Winstanley concluded in words reminiscent of Andreae's: "In all these five Fountains here is Knowledg in the practice, and it is good."

From this practical union of workers in raw metals and minerals and those in related industries, both authors expected useful inventions. Andreae's factories were "the place for originality," and Winstanley wanted to make sure that in the managing of any trade, no young person would be discouraged

> in his invention, for if any man desire to make a new tryall of his skil in any Trade or Science, the Overseers shall not hinder him, but incourage him therein; that so the Spirit of knowledge may have his full growth in man, to finde out the secret in every Art.[34]

Both authors required, however, that the material needs of society be met before the worker was allowed to indulge his creativity in new gadgets. Only after basic requirements were satisfied could Christianopolitans "give play to inventive genius." Winstanley echoed this sentiment when he said that men must be "sure of food and raiment" before their "reason will be ripe, and ready to dive into the secrets of Creation . . . for fear of want, and care to pay Rent to Task-masters, hath hindred many rare Inventions." So they set the stage for the "play" of the inventive mind, and the worker has been given the optimum conditions in which to produce "useful mechanics."

The concern of Winstanley and Andreae for the rich resources lying hidden in the earth's depths was shared by the other utopists. In Astragon's House, Davenant pictured men hurrying from mines with materials for the hot furnaces. Cowley's *Proposition* provided for the study of minerals, and Bacon's furnaces operated constantly at white heat temperature to produce artificial metals.

Extracting industries were essential to England's economic growth in the 17th century, and certain technological problems associated with their development, such as the difficulty of raising ore to the surface, the presence of water in the mines, and the insufficient supply of air, challenged many inventors.[35] Robert Boyle published a treatise on *Inquiries touching Mines;* and a strange disciple of Bacon's, Thomas Bushell, who claimed that he had learned the theory of mining from the Lord Chancellor, finally obtained a parliamentary concession to reopen the Mendip mines. Professor Merton attributes part of the "inventive fever" rampant in the century to the great importance of this particular area of development to the economy. Utopists thus reflected the real concern of England. Contrary to their approach in other instances, they often stopped short of projecting final results, concentrating their efforts on describing instruments and means that could relieve the immediate difficulties.

R. H. described industrial inventions more minutely than his fellow utopists but, after all, he built on the Bensalem that had already perfected mechanical arts to a high degree and could make "stuffs" like linen, silks, tissues, excellent dyes, and "dainty works of feathers of wonderful lustre." Continuing these accomplishments, R. H. took full advantage of ancient as well as modern European inventions. The rarities are explained by a member of Salomon's House:

> Here is also, said he, that ducktile glass, which *Faber*, the Inventour thereof, first presented to *Tiberius Caesar*, which is so plyable that it is not easily to be broken, yeilding to the stroak of the hammer like silver or Iron, and which, though we dayly make of the same, we preserve as a sacred Relique in memory of the Inventour. . . .
>
> Then out of a little box he produced some of that powder, which he called *Expulsative*, ten grains of which mixt (said he) with half the ordinary quantity of Gunpowder for a charg, shall send the bullet as far agin out of a Canon. . . .

> These two lamps, which you here behold (said he) shinning in these two large and close stopt vialls, are of *Incumbustible oyl*, which (so as no air comes to it) will never be extinct: the oyl being composed of a bituminous liquor and that pitchy *Naptha*, which flows out of a kind of brimstone-lime near Babylon. . . .
>
> Then he showed me ordinary Ice, petrifyed and so hardened by art (he said) that it was as usefull as ordinary glasse or Christal, though not so transparent, and which no small fire should thaw.[36]

On the same day that R. H. visited Bensalem, an outstanding inventor was being honored because he had found a "way of making Linnen cloath, and consequently paper of *Asbestinum* or *Linum vivum* that fire shall not consume the writing (which paper is called *Salamandrian*) by the help of some mineral powders and the Spirit of Vitriol."[37] The invention was not so unique as it might appear to the reader who encounters it for the first time. John Wilkins, in his *Mathematicall Magick*, which is a valuable summary of mechanical discoveries, attributed to the ancients the use of "*linum vivum*, or *asbestinum*," which they used to make clothing indestructible by fire.

> The bodies of the ancient Kings were wrapped in such garments when they were put in the funerall pile, that their ashes might bee therein preserved, without the mixture of any other. The materials of them were not made from any hearb or vegetable, as other textils, but from a stone called *Amianthus*.[38]

Francis Bacon had considered the value of "salamander's wool" in making a wick for a candle that would burn perpetually, and he concluded that there may be such a material, "a kind of mineral, which whiteneth also in the burning, and consumeth not."[39]

Together with Wilkins and Bacon, R. H. may have bordered on the domain belonging to miracles in this instance, but he also introduced fundamental inventions such as the fire engine and an instrument to write two copies of a document at once, which reminds one of Sir William Petty's similar tool for double writing. With the invention of such devices, utopia had entered the era of technological advance. We are apt to smile at the thought of salamander's wool for a candlewick, noncombustible linen cloth, or ice so petrified that it served for ordinary drinking glasses, and say all was supernaturally possible in utopia. But these products were typical of those being studied under Royal Society auspices, and there was no thought of indulging in utopian illusions. In fact, when such accusations were made by their opponents, they loudly repudiated them and defended the practical aspects of their program.

The list of works in progress that Sprat appended to his *History* gives an adequate view of their interests and the stage of their thinking: papers were written on the weights of bodies increased in the fire, experiments with a stone called "Oculus Mundi" and with the recoiling of guns, histories of the making of saltpeter, gunpowder, dyes (Petty's paper), and the "Generation and Ordering of Greenoysters, commonly called Colchester-Oysters." Nearly every paper attached to the *History* was concerned with technological, and usually commercial, problems.[40] Within these subject areas were some wild suggestions dealing with the feeding of carp in the air, the changing of gold into silver, and the evidence that a spider is not enchanted by a "Circle of *Unicorns-horn* or *Irish earth*, laid round about it." Regardless of the extremity of some proposals, the whole effort signified exploration. And already the Society had perfected many new instruments. They had presented "Several new kinds of *Pendulum Watches* for the Pocket, wherein the motion is regulated, by Springs, or Weights, or Loadstones, or Flies moving very exactly regular;" they had "a very exact pair of Scales, for trying a number of Magnetical Experiments," a new hearing aid, an instrument for finding the velocity of swimming bodies, new spectacles for seeing under water, and so on the long list went.

The Royal Society, too, had its informal ambassadors and sent extensive inquiries abroad—to places like the East Indies, China, "Tenariff or any high mountain," Barbary, Morocco, Iceland, and Greenland—to collect information for its growing library and museum. Sprat's conclusion to his *History* summarized the areas being researched for their great Natural History:

> what are to be taken notice of towards a perfect History of the Air, the Atmosphere, and Weather: what is to be observed in the production, growth, advancing, or transforming of Vegetables: what particulars are requisite, for collecting a compleat History of the Agriculture, which is us'd in several parts of this Nation.
>
> They have prescrib'd exact Inquiries, and given punctual Advice for the tryal of Experiments of rarefaction, refraction, and condensation: concerning the cause, and manner of the Petrifaction of Wood: of the Loadstone: of the Parts of Anatomy, that are yet imperfect: of Injections into the Blood of Animals; and Transfusing the blood of one Animal into another: of Currents: of the ebbing, and flowing of the Sea: of the kinds, and manner of the feeding of Oysters: of the Wonders, and Curiosities observable in deep Mines.[41]

Here are the utopists' major interests as well, and we could be reading any one of them. They were directly in line with the experimental thought and attempts of their own day, though occasionally they projected further and wisely refrained from describing results in too great detail. In comparison with their notions for industrial products, their dreams for improved commerce and communications have been exonerated from the criticism of fantasy largely because modern civilization has actually realized their dreams. Yet our subsequent progress does not diminish the imaginative height of these inventions for the period in which they were proposed.

Commerce and Communication

As the wagon in the Solarian's field of grain operated by "wheels within wheels" and determined its own direction regardless of which way the wind blew, so the utopists pictured other unique conveyances for travel on land, sea, and air. That same "ingenious Boniger," to whom R. H. gave credit for the remarkable machine that moved three ploughs at once, was the man who

> first gave the vigorous motion to the ship, that by the help of an artificiall *primum movens* within it, and but one man to move the same engin (which is placed on the side of the Vessel) it sailes without help of oares, in the greatest calm, and sometimes against wind and tide. . . . The same man likewise (said he) invented the flying chariots to be born up in the air.[42]

This vessel hardly seems dependable in its ability to go against the wind and tide but, like Campanella's, it could on occasion move against opposition. Campanella added later the explanation, as he did concerning his wagon that moved by sails, that this was due to a marvelous contrivance. And even in the first report of the *City of the Sun*, Campanella said they possessed the art of flying and had found the most intricate way for a warrior on horseback to handle the reins with his feet, thus freeing the hands for effective aim in the fight.[43] Campanella arranged for movement on the water, in the air, and on the ground.

Bacon's society, too, was completely mobile: "We imitate also flights of birds; we have some degrees of flying in the air; we have ships and boats for going under water, and brooking of seas; also swimming-girdles and supporters." R. H. enlarged on Bacon's underwater boat, showing how the Bensalemites had now found means for "making a little vessel to swim under water undiscovered, to blow up ships, bridges and houses." It was obviously a deadly little instrument.

Lest the reader accord R. H. too high a degree of creativity, he must know once more that the Royal Society, too, was fascinated with underwater devices, ranging from a diving bell in which a man might go to great depths, to Hooke's proposal of a "full-fledged submarine which would move as fast as a wherry on the Thames!"[44] John Wilkins was also studying the possibility of building an "ark for submarine Navigation," and he further claimed that Mersenne in France was interested in his project and that Cornelius Dreble in England was actively experimenting with underwater travel.[45] England's leading scientists were keeping up with the imaginative utopists in the age of submarines, airplanes, and river boats.

For land travel, R. H. provided a curious mixture of old and new and freakish means:

> Where the streets are even, we use great Mastiff dogs (of which here we breed many) to draw up and down the streets things upon sledges, made low on purpose, and running on four little wheels. By this means one stout dog that is fed with little or no charge, shall carry or draw as much as any three men.[46]

The ox was retained for the plough or wagon, though the horizontal sail contrivance was also in use; and horses were bred in "Maritim parts, that by looking on the Sea they may acquire more fierceness, and become more emboldned for field service."

By far the greatest new invention for land travel was the sailing coach of "*Simon Stevinius*,[47] that excellent inventour of Geometrical engins and proportions." For all the citizens of Bensalem to admire, a sculptor had portrayed Stevin's likeness "sitting himself in a coach of black marble that seemed to travail without horses." Stevin's chariot was a reality and had actually attracted considerable attention after its exhibition in 1606. Marjorie Nicolson relates that it was

> a great carriage, operated by sails, designed to carry twenty-six passengers. One of the first horseless vehicles, it went at seeming incredible speed over the flat, hard sands of the Dutch seacoast. "In two hours space it would pass from Sceveling to Putton, which are distant from one another about 14 *Horaria Milliaria*, that is, more than two and forty miles." One who had made the passage reported a delightful trip; no matter how strong the wind, the passengers did not feel it since they traveled at the same speed as the wind.[48]

Hugo Grotius spread the fame of the "sailing-chariot" in a poem written shortly after the spectacular performance,[49] and John Wilkins was most interested in the invention. He described in his *Mathematicall Magick* (1648) the "sailing Chariot, that may without horses be driven on the land by the wind, as Ships are on the Sea,"[50] and he proposed improvements for Stevin's design by making the sails movable. R. H.'s utopians were quite abreast of modern developments in honoring Stevin for his invention of the horseless carriage. Problems of motion have engaged the attention of the human race for many centuries, but at no time, unless perhaps our own with its challenge of speed and space travel, has mobility been more important than in this age immediately preceding the steam engine.[51]

The question of motion was not confined to traveling. Bacon saw the general need for continuous power and for increasing the speed and control of motion; so this was a prominent area for research at Salomon's House:

> We have likewise violent streams and cataracts, which serve us for many motions: and likewise engines for multiplying and enforcing of winds, to set also on going divers motions.
>
> . . .
>
> We have also engine-houses, where are prepared engines and instruments for all sorts of motions. There we imitate and practise to make swifter motions than any you have, either out of your muskets or any engine that you have; and to make them and multiply them more easily, and with small force, by wheels and other means. . . . We have divers curious clocks, and other like motions of return, and some perpetual motions. . . . We have also a great number of other various motions, strange for equality, fineness, and subtilty.[52]

R. H. supplied the public recognition that was due Archimedes on whose principles of mechanics both Stevin and Galileo had built. Clever workmen had reproduced Archimedes' "silver Spheare of heaven," showing the orderly courses of planets, sun, moon, and fixed stars by building an "artificial Engine within, moving each wheel and sphear to true and exact distance of time and proportion of figure."

Motion in utopias usually remained a practical problem and not one of research in dynamics. Correspondingly, the lodestone was not so much an object for laboratory experiments in magnetism as it was the means for improved navigation. Several utopists summarized its value briefly, as Bacon did for the society of *New Atlantis*. It was simply a rare stone "of prodigious virtue." But others, like Davenant and R. H., projected its uses to solve a problem that was a knotty one in their era.

The solution of questions in navigation, which included the provision of strong timber for vessels and the preservation of foods for the crew on long journeys, as we saw earlier, was essential to the growth of a nation fast becoming a maritime power, the empire on which the sun would never set. The Royal Society was working hard on ideas for the navy, and Petty had built a double-bottomed boat that met a harsh fate but was valiantly defended in Sprat's *History*. Still the inability to determine accurately the longitude at sea was a most serious handicap. Charles II expressed genuine concern with this problem, and interest in improved navigation led to the construction of the Greenwich observatory later in the century. Christian Huygens, working on the pendulum clock, was primarily concerned with its value to the longitude question, and Newton's work on lunar theory was partly focused on this problem.[53]

On the day of Gondibert's tour through the house of wonders, Astragon had learned the secret, but it came in a moment of enlightened vision and was to be withheld until man improved his corrupt state and returned to religion.

> For this effectual day his art reveal'd,
> What has so oft made Nature's Spies to pine,
> The loadstones mystic use, so long conceal'd
> In close alliance with the courser mine.
>
> And this, in sleepy Vision, he was bid
> To Register in Characters unknown;
> Which Heav'n will have from Navigators hid,
> Till Saturne's walk be Twenty Circuits grown.
>
> . . .
>
> Religion then (whose Age this World upbraids,
> As scorn'd deformitie) will thither steer;
> Serv'd at fit distance by the Arts, her Maids;
> Which grow too bold, when they attend too near.
>
> . . .
>
> Till then, sad Pilots must be often lost,
> Whilst from the Ocean's dreaded Face they shrink;
> And seeking safety near the cous'ning Coast,
> With windes surpriz'd, by Rocky ambush sink.
>
> . . .
>
> Then (sure of either Pole) they will with pride,
> In ev'ry storm, salute this constant Stone!

And scorn that Star, which ev'ry cloud could hide;
The Seamen's spark! which soon, as seen, is gone!

'Tis sung, the Ocean shall his Bonds untie,
And Earth in half a Globe be pent no more;
Typhis shall saile, till *Thule* he descry,
But a domestick step to distant Shore![54]

R. H. also prized the "truely pretious stone" because "by its charitable direction it not only ciments the divided World into one body politic, maintaining trade and society with the remotest parts and nations, but is in many other things of rare use and service."

The magical properties of the lodestone had been known many years: Ralph Hythloday had gained favor among More's Utopians by teaching them its uses, and they no longer feared to venture forth on the sea. Before they knew its powers, sailing trips had been restricted to the summertime. Davenant and R. H. did not go much further than typical poetic imagery in describing the stone, but they emphasized its potential value in guiding men to the unknown parts of the world: "And Earth in half a Globe be pent no more."

In the experimental laboratories of 17th-century utopias, in fields and factories, man's efforts were concentrated on the uses of new discoveries. No time was lost in applying whatever was learned to increase productivity in agriculture, industry, and commerce, and to make man's life longer, healthier, and happier. Though work was directed toward both light and fruit, the fruit was of prime importance and the measure of progress. Utopians were in the atmosphere of applied science and technology. They were not thinking about pure knowledge or theory any more than most of us do when the rocket that carries the heaviest pay load rises aloft into orbit.

Man's control of nature went a long way in the fanciful visions that were plumbing Newton's "great ocean of truth." They may have floundered a bit in the depths, but with imaginative foresight, the utopists projected their fictitious citizen into a mobile age of airplanes, steamboats, submarines, and at least "sailing coaches" on land. He talked by telephone, had his fruits out of season and his crops protected from their natural enemies. He controlled power sources for industrial plants, and he had invented instruments to magnify considerably his sensory abilities. Living in today's world, we can no longer call this "utopian fiction." Only the essential ingredient in many instances—the exact means of how to do it—was missing, and it has taken us more than two centuries to fill this important gap between ideas and reality.

IX

UTOPIAS IN PERSPECTIVE

THE MODERN UTOPIA that describes the automatic life of man in a mechanistic world is most familiar to the contemporary reader. George Orwell in *Nineteen Eighty-Four* portrays man's horrible existence in an all-powerful state that has developed means of communication and control to such an extent that there is no place—even in man's mind—where privacy is possible. The omnipotence of party power over thought renders art, literature, and science sterile. No empirical or questioning habits of thought are permitted, and the official language of Oceana, "Newspeak," has no word for "science." What remains of science is devoted to the art of killing and to the control of man. This world, as an Inner Party member explains, "is the exact opposite of the stupid hedonistic Utopias that the old reformers imagined."[1]

Aldous Huxley expresses a similar concern in *Brave New World* but without the grim emphasis on tyranny and cruelty. Huxley centers attention on scientific power applied to the control of man for what is considered to be his own happiness. To achieve this goal, the human being must be conditioned to remain content in a completely stabilized society.

Beneath his satire, Huxley's affirmative plea is that science and technology should be used and controlled *by* man; man should not be enslaved by them. He states in the foreword to the edition of 1955: "The theme of *Brave New World* is not the advancement of science as such, it is the advancement of science as it affects human individuals. The triumphs of physics, chemistry and engineering are tacitly taken for granted. The only scientific advances to be specifically described are those involving the application to human beings of the results of future research in biology, physiology and psychology."[2]

Both works imply a warning that the individual must proclaim his rights and wield his rational powers lest he be controlled by the new knowledge in the hands of a few leaders. Huxley's symbol of power is a benign force in the person of Mustapha Mond, Resident World Controller for Western Europe; Orwell's Inner Party member, O'Brien, signifies the straight worship of power for its own sake. Nevertheless, the results are similar. Man has lost the freedom to direct his life—even to think his own thoughts. He is the pawn of society. If a cog slips in the social machine so that an individual feels the strain of his existence, Orwell administers Victory Gin and Huxley prescribes soma[3] capsules to raise the spirits. We need not console ourselves with the idea that man at least needs tranquilizers in order to accept the status of a robot. To these two modern utopists science is taking over the reins of control and becoming all-powerful to the detriment of man. H. G. Wells lightens the effect, giving the gadget-age an amusing twist, but with it, as in the *Story of the Days to Come*, there remains the horror of life in a glass-covered London where the hypnotist and doctor are realizing their goal of control over people's lives.

Yet, we have seen the time when science offered only the greatest hope for mankind. Science was to be the fountainhead of blessings and benefits for Andreae's people in *Christianopolis*, Bacon's citizens in *New Atlantis*, and Campanella's society in the *City of the Sun*. These 17th-century utopists stand at the beginning of a line of thought for which Orwell and Huxley demonstrate the end. The result of the progress Bacon anticipated is man the automaton. We may well ask what happened to Bacon's dream. Certainly this ominous conclusion was never his intent. His was still a unified world of values in which there was a larger design than the floor plan of a laboratory. It is true that he banished religious doctrine from the research center and saw man as the creator of his own better world by means of the arts and sciences, but the whole work undertaken was to be a return to man's original perfection before the fall when he possessed complete knowledge. Full responsibility rested directly on man, protected by his reason and values. At times, Bacon sensed the dangers inherent if the powers he saw emerging were not effectively managed. The Janus aspects of the new knowledge he feared when he commented that the technological arts "have an ambiguous or double use, and serve as well to promote as to prevent mischief and destruction, so that their virtue almost destroys or unwinds itself."[4] Again, recognizing the possible misuse of power, he answered his own fears:

> Lastly, let none be alarmed at the objection of the arts and sciences becoming depraved to malevolent or luxurious purposes and the like, for the same can be said of every worldly good; talent, courage, strength, beauty, riches, light itself, and the rest. Only let mankind regain their rights over nature, assigned to them by the gift of God, and obtain that power, *whose exercise will be governed by right reason and true religion.*[5]

So it was on this basis that he proceeded: power was to be controlled by man's right reason and religion.

Writing before him, Andreae and Campanella did not even entertain such suspicions. There was no question about who was in control or for what purposes. In *Christianopolis* the scientist was third man on the scale of authority; representatives of religion and justice preceded him. Although the network of laboratories was installed, medals were given only for virtue, and the greatest reward was pleasing God. Pictures and statues of famous men "with their manly and ingenious deeds" were displayed everywhere in town as incentives "to youth for striving to imitate their virtue."

Campanella's Solarians agreed in part with Andreae's citizens, retaining three collateral princes—Power, Wisdom, and Love—as the ruling council under Metaphysician, and placing science under the jurisdiction of Wisdom; but when they designated areas of knowledge necessary for the high ruler, science ranked as an outstanding requirement:

> No one can be elected *Sun* [Metaphysician] unless he knows the history of all peoples—their customs, rites, and govern-

> ments—and the inventors of all the arts and laws. He must moreover know all the mechanical arts. . . . In addition, he must know the mathematical, physical, and astrological sciences too.[6]

These seemed odd specifications for a ruler to the visiting Hospitaller who asked how anyone giving "his attention to the sciences" could know how to rule. He is answered that "no one can master so many sciences unless he has a ready talent for all things," and "therefore, such a person is always most able to rule." It was a general endorsement of learning with physical and applied science included for the leader of government and society in the *City of the Sun.* Furthermore, the position of central attention, the innermost wall of the city, which showed all the mechanical arts with their inventors, had on its other side "all the founders of laws, of sciences, and of weapons." This assemblage proved to be an assortment of personages including Moses, Osiris, Jupiter, Mercury, Mahomet, with Christ and the 12 Apostles in the most exalted place. The collection itself and its placement are evidence of Campanella's most unified world of religion and knowledge. Both Andreae and Campanella gave the explorer in knowledge a high position in utopia, higher than the majority of society actually awarded at the time, but in neither case was the scientist as such in command.

After the *New Atlantis* appeared, there was no further doubt where the scientists belonged in utopian society. In unequivocal terms (assuming his social responsibility) he was enthroned at the top of the social hierarchy, well ahead of contemporary thought and first in the utopian tradition. When the hall of fame was designed for Salomon's House, there was no confusion concerning the type of hero to be honored. Years before in an unpublished work, "On the Interpretation of Nature" (1603), Bacon had compared the value of the inventors' work with that of heroes and lawgivers, whose contribution lasted only a short time, while the "work of the Inventor . . . is felt everywhere and lasts forever."[7] He reiterated the same point exactly in the *Novum Organum* in 1620.[8] Hence we find that no hero or lawgiver has slipped into Salomon's museum devoted to inventors.

"Two very long and fair galleries" were dedicated: one for a display of patterns and samples of rare inventions, and the other for statues of eminent inventors.

> There we have the statua of your Columbus, that discovered the West Indies; also the inventor of ships: your monk that was the inventor of ordnance and of gunpowder: the inventor of music: the inventor of letters: the inventor of printing: the inventor of observations of astronomy: the inventor of works in metal: the inventor of glass: the inventor of silk of the worm: the inventor of wine: the inventor of corn and bread: the inventor of sugars: and all these by more certain tradition than you have. Then have we divers inventors of our own, of excellent works.[9]

Statues were elaborately made of brass, marble, and touchstone, "some of cedar and special woods gilt and adorned," others of iron, silver, and gold. "For upon every invention of value, we erect a statua to the inventor, and give him a liberal and honourable reward."

Monetary incentive was added to stimulate the flow of inventions. Reward was no longer a matter of pleasing God; it had become a matter of ducats and statues. R. H. stipulated that reward should be proportionate to the worth of the invention and the merit of the person, and one inventor received 5000 gold ducats for a new product. A reward system to advance science and knowledge was after this time a standard feature of scientific utopias. For Macarians, it was ordered that

> all such as shall be able to demonstrate any experiment, for the health or wealth of men, are honourably rewarded at the publick charge; by which their skill in husbandry, physick, and surgery, is most excellent.[10]

In Gerrard Winstanley's society, no man was to be richer than another, but "he who finds out any secret in nature shall have a title of honour given him, though he be a young man." Sir William Petty specified rewards and honors, and Cowley's governors for the ideal college preferred practical inventors for appointment to the faculty.

> If any learned Person within his Majestie's Dominions discover or eminently improve any useful kind of knowledge, he may upon that ground, for his reward and the encouragement of others, be preferr'd, if he pretend to the place, before any body else.[11]

Cowley increased the incentive later in his account, saying that

> if any one be Author of an Invention that may bring in profit . . . and besides, if the thing be very considerable, his Statue or Picture with an Elogy under it, shall be placed in the Gallery, and made a Denison of that Corporation of famous men.[12]

Thus the resourceful inventor, whose name was synonymous with public good, took his place in the gallery of honor with those who had given new knowledge to the world—"as Printing, Guns, *America*, &, and of late in Anatomy, the Circulation of the Blood, the Milky Veins, and such like discoveries in any Art."

Francis Bacon, not stopping with monetary rewards and effigies in galleries, awarded the scientist the position of aristocrat, if not ruler and high priest. Commoners were seldom privileged to see the Father of Salomon's House. Twelve long years had elapsed since the great man had deigned to make an appearance, and, when he came, "there was never any army had their men stand in better battle-array than the people stood." Those in windows also remained "as if they had been placed," while the impressive chariot bearing the Father proceeded in stately fashion through the streets of New Atlantis. Footmen and horses attended the chariot, preceded by 50 young men dressed in white satin, silk, and blue velvet. Other followers were dressed predominantly in this color scheme with varied hues flashing in the silk carpet beneath the Father's feet. On top of the decorative chariot was a "sun of gold, radiant" and a small golden cherub with wings outspread. Two men, one bearing a bishop's staff and the other "a pastoral staff like a sheep-hook," walked before the chariot while "all the officers and principals of the Companies of the City" followed behind it. As the procession moved along in its splendor, the Father held up his hand blessing the people. Highest esteem was due the director of "the noblest foundation . . . that ever was upon the earth." And a comparison of the honor bestowed on him with that granted to Tirsan at the Feast of the Family readily reveals that there was no competitor for the Father of Salomon's House. Bacon's quill, regardless of his thoughtful warnings, spoke more frequently and more confidently for the new approach; the lists of projects to be investigated and the power to come with knowledge was the message heard by others, not the cautionary tone. His followers remembered primarily that Bacon conceived of Salomon as one

> that whilst he flourished in the possession of his empire, in wealth, in the magnificence of his works, in his court, his household, his fleet, the splendor of his name, and the most unbounded admiration of mankind, he still placed his glory in none of these, but declared that it is the glory of God to conceal a thing, but the glory of a king to search it out.[13]

The search was under way with high stakes promised: this was what psychologically attracted and inflamed the joiners who saw Salomon's House as the palace of the future.

R. H. apparently realized that his predecessor had given the greatest luxury, dignity, and wisdom to the Father of Salomon's House, so he enhanced the prestige of the ordinary scientist in Bensalem. On the occasion he described, Verdugo, the inventor of Salamandrian paper, received the accolades, a purse of gold, and an annual pension for life. Verdugo rode in the chariot this time in a parade designed in classical fashion. He wore a laurel wreath on his brow and a mantle over one shoulder. A youth impersonating Minerva, the goddess of invention, carried a roll of the new paper fired at both ends. It was the custom in these ceremonies for the emblem to reflect the invention being honored. Similarly, society had instituted a "solid kind of heraldry" showing the person's virtue or achievement. For example, if a man excelled in geometry, his emblem was the astrolabe; in arithmetic, a "Table of Cyphers;" in physic, a urinal; and the more or less usual marks, a turtle dove for chastity and the sun for charity, were retained.

The entertainment was most elaborate, including a pastoral interlude, and Minerva's crown engraved with great inventors' names was placed on Verdugo's head. Yet the ritual was also functional and educational. A speech of commendation was delivered on ingenuity and learning in general, with a particular encomium on the invention and a list of its benefits to all. The inventor replied by explaining the process of his invention, and then recorded the exact formula for producing the excellent paper. R. H. interpolates that this was not only to prevent monopoly and stimulate others but to insure that the invention will not perish with the author.[14] At the end of the program, the Father of Salomon's House entered the invention in a book made of the new Salamandrian paper. The most heralded event in utopian society was a new discovery or technological improvement of a useful sort. Moreover, research and production were completely supported by governmental funds, which was to be expected because the benefits were shared by all citizens.

It was the pre-eminent position being given the experimental scientists that annoyed Samuel Gott. But citizens of his *Nova Solyma* constituted a minority opinion at midcentury when they registered scorn for those who "make of *human* art an idol." Others did not agree at all with their statement that "short-sighted, foolish mortals are ready to give almost divine honours to such discoveries [printing press, gunpowder, and the mariner's compass], shallow though they be." Such discoveries were far from shallow to the majority of utopians who have happily and enthusiastically given their praise to the scientist, inventor, or discoverer. The spotlight had shifted from the philosopher-king of Plato and Thomas More to the scientist-king. Seventeenth-century utopians lived in the secular world of affairs and utilitarian values. They were a long way from the freely flowing plenitude taken for granted in ancient idealistic utopias. Theirs was planned, the result of effort and education, so each day they saw their lives improved materially and their control of nature extended. They enjoyed the effects of their increased knowledge in very practical ways: by having their fruits when they pleased, regardless of nature's season, and by traveling with speed on the ocean, regardless of the tide and winds. Such accomplishments were evidence of their victory over the elements and, consequently, their measure of progress. They were quick to use the power of the new knowledge for their own benefit, for utilitarian goals; and thus, they lived in the age of technology and applied science.

Exhilarated by unlimited possibilities for the future, utopians rejoiced that they had found the practical formula to realize their vision. Their formula, of course, was the right method, which they instituted in both classroom and laboratory. From great fact collections—the epitome of the whole world—they will sift out the new truths. Although workers were occasionally assigned specific tasks in the scheme leading to higher truths, the experimental process was not precisely stated. One principle, however, was certain: nothing would be accepted as truth unless it had been actually observed. In the reaction against Aristotelianism, utopians believed only what could be proved to their senses, and they built inductively on this basis alone.

Often workers were not too sure about the particular purpose of their experiment. Andreae's Christianopolitans, among the most progressive, still tried to determine motion as it was derived from the Primum Mobile. Furthermore, they directed attention to the qualities, habits, and uses of the object under examination, pondering such questions as: "For what purpose living animals and plants exist, of what use metals are, and especially what the soul, that spark of divinity within us, accomplishes." Andreae's community was out of step with Galileo and other scientists, who were concentrating on the measurable and predictable elements of phenomena. They were searching for efficient causes that could be predicted to obtain certain results, and from their work came the analyses that uncovered those irrefutable, stubborn facts of which universal laws are formulated. New utopians, however, generally agreed with Christianopolitans and went on with their rather vague, but promising, collection and classification of facts. Their broadside approach cannot be critically held against them in evaluating their contribution, because it was the amorphous period of scientific development when lines were not clearly drawn, methods often vague, and definitions far from standardized.

Not the details, but the "idea" of science challenged the utopists. They were creative, albeit literary, scientists, embracing the new facts with excitement and imaginatively projecting them into man's life. Utopists were powerful publicists in extending the scientific doctrine and in gaining its acceptance. To do this effectively required the removal of older doctrines that had prescribed man's thinking and activity. Hence, the utopists answered the controlling authority of church doctrine by asserting that science was perfectly compatible with God's purpose no matter what the literal Scripture said. In some instances, they decisively separated the two areas of thought and refused to admit matters of faith into the laboratory. Finally Joseph Glanvill, the last of the group, ushered in the age of reason, denying the supernatural, insisting on a rational God, and a reasonable, even naturalistic religion.

For similar reasons, utopians rejected the despotic rule of Aristotelian thought and the dominance of ancient authorities in established educational institutions. They proposed, instead, a progressive education for youth based on the scientific method of observation. Children learned visually from the object—not by rote memory from the book and lecture. All graduated with a vocational skill that could earn them a living. Man was a tool to be polished and sharpened to carry on the laboratory work and to fulfill the practical needs of a material society.

With these answers to existing authority, utopists cleared the way for new worlds of thought. They established with Glanvill the right to question skeptically and the liberty to theorize, thus setting the stage for the spectacular rise of modern science. While they purged the intellectual atmosphere of its discouraging and retarding elements, they optimistically described the new scientific, utilitarian world of the future, which they believed was imminent and wholly possible in their own countries.

To show their contemporaries the alluring goal, these authors chose the literary genre of utopia as the most fitting for their purpose. In the freedom of fiction they could describe the ideal society realistically. Moreover, in their descriptions, the genre itself underwent interesting changes. The traditional element of authenticity, which had been the writer's amusing effort to make his story "seem" real, was now either abolished altogether by presenting the utopia in a straightforward essay style, or directed to the serious business of convincing the reader that the ideal state could be his if he would work to make it so. The new utopists were reformers presenting their philosophic ideal, but as a realistic possibility, not as a statement for contemplation but a plan for action. We observed that several authors actually tried to start their ideal societies. Their use of the utopian genre was as a means to advertise the new product—the scientific world they envisaged.

Both the form and content of the traditional utopia have been basically revised, and the feeling of a utopian dream is gone. The new utopists no longer portray the final, absolute ideal. That is impossible because they are committed to the evolutionary principle: their knowledge of the universe was to grow, accumulate, and change with counterfacts discovered. Evolutionary doctrine in the sense of man's development as a higher organism was, of course, not suspected, but evolutionary logic, starting from the fact of thinking as inquiring and building inductively through a process of selection, rejection, and survival is an inherent part of the new method. Furthermore, Bacon was as evolutionary in logic as John Dewey in believing that the successful idea is a true idea. When a process becomes the central issue of concern, the static ideal is completely overthrown, and the dynamic pattern of evolutionary progress takes its place. There is no limit to what the new approach to knowledge may reveal, and life can be constantly improved.

This belief our utopists certainly held, although they expected the mechanistic universe would be the final natural law. Here they and their fellow scientists have been proved wrong by theories of relativity and indeterminacy. Human nature searches for the infinite but is more comfortable with the finite and so often accepts it as "truth" too quickly. Even when free to dream, we welcome the security of limits; still their limits were wide indeed for their day.

For their vision they are indebted to many long lines of thought and belief—the rational approach of the Greeks and their admiration for the powerful ability within a man; the dynamism of the Hebrews both in the striving for heaven and in the moving view of heaven itself; the concept of a reasonable God who would not act upon a whim in an unpredictable fashion and therefore did not create a quixotic universe. All such concepts seen in early heavenly utopias had joined together and created the elementary environment conducive to ideas of progress and growth in knowledge. From this premise, man could choose the direction for his energies and hopefully pursue it. On the far boundaries of his known world he created ideal societies to be envied and emulated. To this was added the growing concern for the present and its possibilities, which humanism reintroduced and urged upon those who were intellectually curious and ambitious. More utopias were conceived and progress was guided by economic, political, and educational reforms as well as ethical standards. With the announcement of new revolutionary knowledge came the group of new utopists who would bend all efforts toward its further advancement and application. Their gift was great to the scientist working in the laboratory: he could walk forth from a day's labor and receive acclaim and support from his neighbors who had gained from the interpreters a general vocabulary with words like experiment, method, facts, and finally natural laws.

With all their imaginative faith in the scientific process, 17th-century utopists, however, could no more have imagined the extreme dominance of science in modern society than Isaac Newton could have realized that he introduced the Age of Satellites. Yet Newton in his *Principia Mathematica* (1687) correctly "diagrammed the launching of a satellite and showed the speed necessary—five miles a second—to keep a body in motion around the earth."[15] Both the utopists and the scientist Newton lived in the formative period of a concept of progress based on the conquest of nature, that is, science.

But somehow this concept has led to the conquest of man, too, in the utopian societies of Orwell and Huxley. It is the very condition that Bacon feared as possible if man did not rationally and religiously direct the new powers. Bacon's hopes and dreams for the beneficial use of science, however, found their fullest expression to date in H. G. Wells' *Modern Utopia* where Salomon's House operates world-wide and sustains over a million men, where federal funds amply support research and creative activities in many areas.[16] Wells saw his expansion of mechanical possibilities foreshadowed in the *New Atlantis* and now, in its modern phase, there is to be an army corps, a multitude of selected men, collaborating in their work. Every university in the world is urgently working for priority in one or another aspect of a problem. Reports of experiments are telegraphed about the world. Great systems of laboratories attached to municipal power stations conduct research "under the most favorable conditions;" and every mine and almost every large industrial establishment does the same. We may, at twilight, suddenly come upon "a thousand men at a thousand glowing desks, a busy specialist press," and they will be "perpetually sifting, criticizing, condensing, and clearing ground for further speculation." Some are concentrating on problems of public locomotion, aeronautic investigations, while others are at work in physiology and sociology. As a result of this intensive effort, utopian research goes like "an eagle's swoop" right at the target.

The method is less clear than Bacon's, although the approach is obviously similar. The literary Wells is not interested in techniques any more than are most utopian dreamers. Like them his concern is for results and benefits. In these he goes well beyond his 17th-century predecessors and into our modern society: mechanical force is coming from nature to the service of man, who is thus emancipated from physical labor (on which other utopias were based) and freed to play and invent, to enjoy his leisure which, in a good moral and intellectual atmosphere, leads to experiments, philosophy, and new departures. Servitude and inferiority are abolished. True freedom with its ultimate significance in individuality is the result. People may be numbers in an extensive registration system in Wells' *Modern Utopia*, but their lives are led in the liberty to develop themselves. Wells states his attitude directly in reference to our world:

> The plain message physical science has for the world at large is this, that were our political and social and moral devices only as well contrived to their ends as a linotype machine, an antiseptic operating plant, or an electric tram-car, there need now at the present moment be no appreciable toil in the world, and only the smallest fraction of the pain, the fear, and the anxiety that now make human life so doubtful in its value. There is more than enough for everyone alive. Science stands, a competent servant, behind her wrangling underbred masters, holding out resources, devices, and remedies they are too stupid to use.[17]

While it is fortunately unlikely that the human equation, politically or socially, will ever be solved so efficiently and effectively as the linotype or tram-car for its purposes, Wells' lament comes to many with understanding and agreement. Throughout nearly three centuries we have developed material goods and the means for making man's life less arduous, more pleasant and healthful. It may be that this goal is insufficient for the wondrous creature that is man. We have in many ways realized the visions of Andreae, Campanella, Bacon, or even Wells, and if the results are in some respects frightful, it is not their dream, but the values of man that are at fault. "Science has not given men more self-control, more kindliness, or more power of discounting their passions in deciding upon a course of action."[18] Science cannot give us the social invention that we must have, not only to keep pace with technology's inventions, but to determine which discoveries we will reject, which we want for the good of society, and how we will adjust the imbalance caused by each major invention. There is no reason to accept all results from laboratories as automatically good or of real value. This is the task of the social scientist, the philosopher, and humanists who deal with values in the life of men. Dennis Gabor, a modern technologist, has summarized the issue and the need:

> It is a sad thought indeed that our civilization has not produced a *New Vision* which could guide us into a new "Golden Age" which has now become physically possible, but only physically. . . . Who is responsible for this tragi-comedy of man frustrated by success? . . . Who has left Mankind without a vision? The predictable part of the future may be a job for electronic predictors but that part of it which is not predictable, which is largely a matter of free human choice, is not the business of machines, nor of scientists . . . but it ought to be, as it was in the great epochs of the past, the prerogative of the inspired humanists.[19]

He concludes: "The future cannot be predicted, but futures can be invented."

Perhaps the new utopists of the late 20th century will recall Plato's concept of progress toward virtue or Sir Thomas More's concept of achievement toward the ideal behavior of man. These ideals will always have validity for the individual, but their power as goals for the progress of a total society or civilization appears to have been eclipsed by more dynamic, exciting concepts, by the idea of worlds of knowledge as yet undiscovered. There is no incompatibility between the two goals: the ethical ideal and the intellectual conquest of the unknown. Without the one, the other is considerably weakened; an Orwell or a Huxley could have predicted correctly. Man dare not forfeit his position as master and director of the child of his own brain. Science has developed at a phenomenal rate and created highly specialized areas of thought that leave the majority of men far behind in understanding. If we are to control this threatening yet beneficent giant, it is imperative that we understand its methods, its fallibility, and its great promise. Thus science may still play the role envisaged by the early utopists once our values and general understanding guide its development and gifts to mankind.

The utopists of the future may well project a world as yet unimagined and it, too, may become reality, just as the 17th-century utopists described a society that we must recognize in several important respects as the world we now inhabit.

> A map of the world that does not include Utopia is not worth even glancing at, for it leaves out the one country at which Humanity is always landing. And when Humanity lands there, it looks out, and, seeing a better country, sets sail. Progress is the realization of Utopias.[20]

NOTES

THE REFERENCES BELOW are not only for documentation but also to indicate where fuller or more specialized treatment may be found. I have given the publisher's name as well as place and date for all works after 1900. I have also given the complete reference the first time a work is referred to in each chapter. Two abbreviations are used: LCL for Loeb Classical Library (texts and translations of Greek and Latin authors published currently by the Harvard University Press) and OED for Oxford English Dictionary. PMLA and ELH are carried as actual titles of these journals.

INTRODUCTION

1. *The Tempest* II.i.158–164. The word "foison" means plenty.

2. The word "utopia" has been interpreted to mean not only u-topia, "no place," but also eu-topia or "good place." Other terms based on topia (place) have more recently been coined. Professor J. Max Patrick, an authority on utopias, uses "dystopia" for the negative of the ideal society. See Glenn Negley and J. Max Patrick, *The Quest for Utopia* (New York: Abelard-Schuman, 1952) for a discussion of terminology and a comprehensive anthology of utopian literature. The word "cacotopia," meaning "bad place," has been used by W. H. Ferry, Michael Harrington, and Frank L. Keegan in their *Conversation: Cacotopias and Utopias* (Santa Barbara, California: Fund for the Republic, 1965). To clarify the meaning further, Constantinos A. Doxiadis recommends "eftopia," the modern Greek pronunciation for the good place, eutopia. See his book, *Between Dystopia and Utopia.* (Hartford, Connecticut: Trinity College Press, 1966), p. 25.

3. *The Epic of Gilgamesh*, trans. and introd. N. K. Sandars

(Baltimore, Maryland: Penguin Books, 1960), introd., p. 13. The four copies are all from Sultantepe, and the text has been translated and published recently by Dr. Oliver Gurney.

4. Benjamin Bickley Rogers, in the introduction to his translation of *The Ecclesiazusae*, in *Aristophanes* (LCL), III, 246, held that Aristophanes heard of Plato's work while writing his play, which would also account for the shift from conservative to revolutionary policies in the drama. J. Van Leeuwen agreed that Aristophanes had Plato in mind. Professor Moses Hadas disagreed, claiming that the *Republic* (which Rogers dates at 394 B.C.) was later, and further that if Aristophanes had been parodying Plato he would have made it clearer, caricaturing him more broadly as was his regular practice elsewhere. He suggested that both men may have drawn from an earlier source that may also have been available to Herodotus in his passages on the community of wives (1.216; 4.172,180).

5. *Lucian, Satirical Sketches*, trans. Paul Turner (Baltimore, Maryland: Penguin Books, 1961), p. 279. This current edition is without book and selection numbers. The quotation is from bk. 2.17.

6. Jean Bodin, *The Six Bookes of a Commonweale*, trans. Richard Knolles (London, 1606), I, 3. In his letter "To the Reader" Knolles reiterates Bodin's point.

7. *An Apology . . . against Smectymnuus*, in *The Works of John Milton*, ed. Frank A. Patterson, *et al.* (New York: Columbia University Press, 1931), III, pt. 1, p. 294.

8. Mark Van Doren, *Shakespeare* (New York: Doubleday, 1953), p. 286.

9. *An Anatomie of the World. The First Anniversary*, in *The Poems of John Donne*, ed. Sir Herbert Grierson (London: Oxford University Press, 1951), ll. 75–78.

10. Lewis Mumford, *The Story of Utopias* (New York: P. Smith, 1941), p. 22. For a more recent discussion, see Mumford, "Utopia, The City and The Machine," *Daedalus* (Spring 1965), pp. 271–292. The issue is devoted to analyses of utopias and has been reprinted together with several new essays in *Utopias and Utopian Thought*, ed. Frank E. Manuel (Boston, Massachusetts: Houghton Mifflin, 1966). The affirmative view of utopian influence is expressed especially in the chapters by Frederik L. Polak, "Utopia and Cultural Renewal," and by Paul Tillich, "Critique and Justification of Utopia," pp. 281–309.

I : THE HEAVENLY PARADISE

1. *Ancient Near Eastern Texts Relating to the Old Testament*, ed. James B. Pritchard (Princeton, New Jersey: Princeton University Press, 1955), 2nd edition, corrected and enlarged. Sumerian myth, the deluge tablet, trans. S. N. Kramer, p. 44. Kramer (note 60) explains "land of crossing" as "perhaps the crossing of the sun immediately upon his rising in the east; the Sumerian word used may also mean 'of rule.' " For the story of Gilgamesh's journey to Dilmun I have drawn on Kramer's translations as follows: Tablet IX of the Nineveh version concerning the lions and the road of the sun (pp. 88–89); Tablet X, which included both the Old Babylonian version concerning Siduri (pp. 89–90) and the Assyrian version concerning the trip across the waters of death (p. 91).

2. *Ibid.*, Tablet XI, trans. Kramer, p. 95. Dilmun was an actual city or land, according to historical cuneiform references, as well as a fabulous place for gods and spirits. Kramer locates it in southwestern Iran and presents his arguments in "Dilmun, the Land of the Living," *Bulletin of The American Schools of Oriental Research*, XCVI (1944), 18-28. P. B. Cornwall, in an article "On the Location of Dilmun" in the same *Bulletin*, CIII (1946), 3–11, argues more convincingly for the island Bahrein. In any case, it seems likely that it was in the Persian Gulf, called by Assyrians the "sea of the rising sun."

3. *Ancient Near Eastern Texts*, p. 38. Kramer explains (note 17) that the *ittidu*-bird is "probably a bird whose cry is a mark of death and desolation." Square brackets indicate restorations; parentheses indicate interpolations for understanding; italics are used for a doubtful translation or to indicate transliteration.

4. *Amos* 9.13–14.

5. *Odyssey*, trans. Samuel Butler (London: A. C. Fifield, 1900), 6.43ff. Subsequent references will be to this translation. Moses Hadas makes this observation concerning Homer's description in his excellent chapter "Blessed Landscapes and Havens," *Hellenistic Culture* (New York: Columbia University Press, 1959), pp. 212–222. From this I have drawn heavily in the present section, but my interpretation is necessarily somewhat different, and the range goes beyond classical and Hellenistic worlds.

6. *Works and Days,* in *Hesiod. The Homeric Hymns and Homerica,* trans. Hugh G. Evelyn-White (LCL), ll. 170–173.

7. *The Odes of Pindar Including the Principal Fragments,* introd. and trans. Sir John Sandys (LCL), Fragments 129 and 130.

8. *Aeneid,* trans. H. R. Fairclough (LCL), 6.637ff.

9. Trans. H. W. and F. G. Fowler (Oxford: Clarendon Press, 1905), bk. 2.5–6, 12–13.

10. Hadas, *Hellenistic Culture,* p. 216.

11. *The Apocalypse of Peter,* trans. and introd. Andrew Rutherford in *The Anti-Nicene Fathers. Translations of the Writings of the Fathers down to* A.D. *325,* ed. Allan Menzies (Grand Rapids, Michigan: William B. Eerdmans, 1951), X, secs. 15–18.

12. *Ibid.,* secs. 8–10.

13. *Ibid.,* introd., p. 142.

14. *Paradise Lost,* in *The Works of John Milton,* ed. Frank A. Patterson, *et al.* (New York: Columbia University Press, 1931), II, pt. 2, 5.426–430. Subsequent references will be to this edition.

15. *Ibid.,* 5.633–638.

16. Harold H. Watts, *The Modern Reader's Guide to the Bible* (New York: Harper, 1959), p. 497. He also mentions here the scarcity of Biblical statements concerning Enoch.

17. *The Book of Enoch the Prophet,* trans. Richard Laurence (1821), corrected and introd. Charles Gill (London, 1892), 14.12. *The Apocalypse of Abraham,* trans. G. H. Box (New York: Macmillan, 1919), also has angry cherubim jealously guarding God's throne.

18. *The Book of Enoch,* 4.

19. *Ibid.,* 31.1–4.

20. *Ibid.,* 59.9. The translation "refrigeration" is misleading. *Refrigerium* has a technological, theological suggestion more like "spiritual serenity."

21. *Ibid.,* 59.12.

22. *Ibid.,* introd., p. 21. Richard Laurence studied Enoch's astrology and came to this conclusion.

23. Hadas, *Hellenistic Culture,* p. 52.

24. They first appear in the Elysian Fields described in the pseudo-Platonic *Axiochus* (371c), which was written probably in the first century B.C.

25. Hadas, *Hellenistic Culture,* p. 78.

26. The description of the Vision of Er is from the *Republic,* in

The Works of Plato, trans. Benjamin Jowett (New York: Tudor, n.d.), 10.614–620.

27. *Iliad*, trans. Samuel Butler (London: Jonathan Cape, 1936), 17.628. See Benjamin Farrington's discussion on this in *Science in Antiquity* (London: Oxford University Press, 1950), pp. 33–50.

28. Farrington, *Science in Antiquity*, p. 37.

29. *The Book of Enoch*, 8.1.

30. *Ibid.*, 68.11–13.

31. *Paradise Lost*, 12.575–582. See Howard Schultz's detailed development of this argument in *Milton and Forbidden Knowledge* (New York: Modern Language Association, 1955).

32. Trans. Andrew Lang, Walter Leaf, and Ernest Myers (London: Macmillan, 1923), 18.416–419.

33. Athenaeus, "Beluae," in *The Deipnosophists*, trans. Charles Burton Gulick (LCL), III, 6.267. Although Crates' fragment is cast as a prophecy, Athenaeus considers it a parody of the Golden Age, as do Arthur O. Lovejoy and George Boas in *Primitivism and Related Ideas in Antiquity* (Baltimore, Maryland: Johns Hopkins Press, 1935). See chap. 2, pp. 23–102, for this and other descriptions of perfect societies. See also David Winston, "Iambulus: A Literary Study in Greek Utopianism" (unpubl. diss., Columbia University, 1956), p. 25.

34. *Satirical Sketches*, trans. Paul Turner (Baltimore, Maryland: Penguin Books, 1961), p. 278. This current translation is without book and section numbers. The quotation is from bk. 2.14.

35. *Paradise Lost*, 5.254–256.

36. Bertrand Russell, "Science," in *Whither Mankind. A Panorama of Modern Civilization*, ed. Charles A. Beard (New York: Longmans Green, 1928), p. 73.

37. Amos was the first of the prophets to extend Jehovah's rule and promise to nations beyond the borders of Israel. The Hebrews' God expressed through Amos (9.7) His concern for all people, including the Negroes of Africa. For a full discussion, see Robert H. Pfeiffer, *Introduction to the Old Testament* (New York: Harper, 1948), pp. 580–584.

38. Joyce Hertzler, *The History of Utopian Thought* (New York: Macmillan, 1923), p. 25.

39. *Works and Days*, in *Hesiod. The Poems and Fragments*, trans. A. W. Mair (Oxford: Clarendon Press, 1908), ll. 110–120.

II : SECULAR VIEWS OF UTOPIA

1. Trans. Moses Hadas in *Three Greek Romances* (New York: Doubleday Anchor, 1953), p. 38.

2. *The Phaenomena of Aratus*, trans. G. R. Mair in *Callimachus and Lycophron. Aratus* (LCL), ll. 96–136.

3. Arthur O. Lovejoy and George Boas, in *Primitivism and Related Ideas in Antiquity* (Baltimore, Maryland: Johns Hopkins Press, 1935), pp. 34–36, discuss developments in the legend and employ the general division of "hard" and "soft" primitivism.

4. *Metamorphoses*, trans. Lovejoy and Boas (*Primitivism*, pp. 46–47), from the text, ed. F. A. Riese (Leipzig, 1872), 1.89–112.

5. *The History of Herodotus*, trans. George Rawlinson (London: Dent, 1945), I, 2.28. Subsequent references will be to this translation and all to vol. I.

6. *Odyssey*, trans. Samuel Butler (London: A. C. Fifield, 1900), 4.85–88.

7. Herodotus, *History*, 3.23.

8. *Ibid.*, 3.18.

9. *Ibid.*, 3.114.

10. *Orlando Furioso*, trans. John Harington (London, 1634), bk. 34, stanzas 49–86. The quotation is from stanza 60.

11. Sir Thomas More, *Utopia and a Dialogue of Comfort* (London: Dent, 1951), p. 79. This edition contains Ralph Robinson's English translation of the *Utopia* (1551) with spelling and punctuation modernized. Subsequent references will be to this edition. The similarity of passages is striking but, of course, the evil of gold had long been seen by many writers. On this see Moses Hadas, "Utopian Sources in Herodotus," *Classical Philology*, XXX (April 1935), 113–121. He cites relevant passages also in Plato (*Laws*, 5.742), Tacitus (*Germania*, 5), and Horace (*Od., III.* 3.49), and suggests the possibility of earlier written accounts, no longer extant, upon which Herodotus may have drawn in the story of Ethiopians and other remote people.

12. More, *Utopia*, p. 81.

13. Trans. Richard Aldington in *The Portable Voltaire* (New York: Viking, 1949), pp. 273, 275.

14. After Homer, who differentiated between the Eastern and Western Ethiopians, Herodotus (7.70) explained that they were

similar except in language and the "character of their hair." The Libyan Ethiopians had woolly hair, and the Eastern branch had straight hair. And Herodotus (4.85) tried to support Homer's account of horns budding quickly on the foreheads of lambkins by suggesting that this was possible in warm climates, while in cold temperatures horns do not come at all or only with difficulty. Aeschylus extended the Ethiopians' area to India, but later they were localized definitely in the country south of Egypt.

15. "Hymn VII to Dionysus," *Hesiod. The Homeric Hymns and Homerica*, trans. Hugh G. Evelyn-White (LCL), l. 28. This is the earliest reference to Hyperboreans, and the context reveals the attitude toward them: "I reckon he is bound for Egypt or for Cyprus or to the Hyperboreans or further still."

16. Herodotus, *History*, 4.32–36. This passage includes mention of Abaris, said to have been a Hyperborean who traveled around the world with his arrow and without eating. See also 4.13–15 for the account of Aristeas of Proconnêsus concerning the Hyperboreans.

17. The location of the Hyperboreans was widely discussed and projected, attesting the seriousness with which many considered these legendary folk. While the far reaches of the north was the general opinion, some were more specific. Hellanicus (450 B.C.) placed them beyond the Rhipean Mountains; Diodorus Siculus (1st century B.C.), recounting what Hecateus of Abdera and "others" had to say about Hyperboreans, locates them on an island beyond the Celtic territory; some place them in northern Asia; and Posidonius Apamensis puts them near the Italian Alps! See Lovejoy and Boas, *Primitivism*, pp. 304–314, for these facts and other detailed accounts of Hyperboreans.

18. "Pythian Ode X," in *The Odes of Pindar*, introd. and trans. Sir John Sandys (LCL), ll. 37–48.

19. *Natural History*, 4.12.89–91.

20. *Concerning the Face Which Appears in the Orb of the Moon*, in *Plutarch's Moralia*, trans. Harold Cherniss and William Helmbold (LCL), XII, 941a–942c.

21. *Ibid.*, 942b.

22. Kepler said in note 2 appended to his *Somnium seu Astronomia Lunari* (Frankfurt, 1634) that he found Plutarch's little book in Graz in 1595, and decided not to use for his work the name of Plutarch's islands, which he located in the Icelandic Sea, but instead chose Iceland itself as the scene for his dream. Plutarch's work and

Lucian's *True History* were background for Kepler's idea to take the reader to the moon where he might then view the "earth, the sun, planets, and fixed stars, and see just how this universe does function." Kepler wished to convince the reader of the truth of the Copernican hypothesis. To the *Somnium* he attached a Latin translation of Plutarch's *De Facie Orbis*. For further information, see Max Caspar's biography *Kepler*, trans. C. Doris Hellman (New York: Abelard-Schuman, 1959), pp. 351–353. See also Joseph Keith Lane, "The Dream or Posthumous Work on Lunar Astronomy by Johann Kepler" (unpubl. master's thesis, Columbia University, 1947), for the English translation including Kepler's notes. Also David Winston, in "Iambulus: A Literary Study in Greek Utopianism" (unpubl. diss., Columbia University, 1956), note, p. 77, refers to Kepler's efforts to identify the islands. Marjorie Nicolson relates the *Somnium* to other literature and thought of the period in "Kepler, the *Somnium*, and John Donne," now reprinted in *Science and Imagination* (Ithaca, New York: Cornell University Press, 1956), pp. 58–79.

23. The Athenians' history is given in two dialogues: *Timaeus*, 20–25, and *Critias*, 109–112.

24. *Critias*, trans. A. E. Taylor in *Plato, The Collected Dialogues*, ed. Edith Hamilton and Huntington Cairns (New York: Pantheon, 1961), 115a–b.

25. Martin Gardner, *Fads and Fallacies in the Name of Science* (New York: Dover, 1957), chap. 14, "Atlantis and Lemuria," pp. 164–172. Mr. Gardner conservatively estimates that a list of "titles on Atlantis, in all languages, which have been published in the present century would run into several thousand."

26. Euhemerus' *Sacred History* (c. 300 B.C.) is preserved in fragmentary form in *Diodorus of Sicily*, trans. C. H. Oldfather (LCL), III, 5.41–46 and 6.1.4–11. References to Panchaea in bk. 6 are directly attributed to Euhemerus, but he is not named as the source for the bk. 5 passage, although it is generally considered to have been drawn from Euhemerus.

27. That is, "Zeus of the three tribes," because the inhabitants of Panchaea derive from three distinct people (*Diodorus of Sicily*, III, 5, note, p. 215). The term "euhemerism" comes from Euhemerus' theory that gods were but deified mortals; hence myth was to be interpreted as an account of historical personages and events.

28. *Diodorus of Sicily*, III, 5.45.3–4.

29. The main source for Iambulus' story is also *Diodorus of Sicily*, trans. C. H. Oldfather (LCL), II, 2.55–60. The work is dated c. 165–150 B.C. on internal evidence. David Winston in "Iambulus," summarizes the tale (pp. 29–38).

30. Gian Battista Ramusio, "Navigatione di Iambolo mercatante antichissimo," in *Delle Navigationi et Viaggi* (Venice, 1550), I, 188–190. Ramusio included Diodorus' excerpt of Iambulus and a short essay concerning it. He believed in the authenticity of Iambulus' account and charted his journey on a map.

31. Winston, "Iambulus," p. 47, discusses this geographical fact as important in dating the life of Iambulus. Eratosthenes (c. 275–194 B.C.) was the first systematic geographer. His greatest work, the *Geographica,* included a system for calculating latitude and longitude which he devised. He also computed with a high degree of accuracy the circumference of the earth.

32. Winston, "Iambulus," pp. 72–73.

33. Moses Hadas, *Hellenistic Culture* (New York: Columbia University Press, 1959), pp. 221–222.

34. *Geography,* trans. Horace Leonard Jones (LCL), I, 1.2.23.

35. *Satirical Sketches,* trans. Paul Turner (Baltimore, Maryland: Penguin Books, 1961), preface, p. 249.

36. See Hadas, *Hellenistic Culture,* especially chap. 6, "Education: Gentlemen, Scribes, Saints," pp. 59–71.

37. *Ibid.*, p. 61.

38. *Ibid.*

39. Trans. Walter Miller (LCL). Subsequent references will be to this translation and cited as Xenophon, *Cyropaedia.*

40. *Ibid.*, I, 1.1.6.

41. Moses Hadas, *A History of Greek Literature* (New York: Columbia University Press, 1950), p. 126.

42. Xenophon, *Cyropaedia,* I, 1.2.1.

43. The words are those of Richard Mulcaster (1530?–1611), an outstanding schoolmaster of the 16th century, quoted by Knox Wilson in "Xenophon in the English Renaissance from Elyot to Holland" (unpubl. diss., New York University, 1948), p. 274.

44. Knox Wilson, "Xenophon," p. 24. The translation did not actually appear until later, after the death of Prince Henry, when Holland's son presented it to Charles I.

45. W. W. Tarn, *Alexander the Great* (Boston: Beacon Press, 1956), p. 113.

46. *Ibid.*, pp. 124–125.

47. *Ibid.*, p. 131.

48. *Ibid.*, p. 147.

49. *Ibid.*, pp. 147–148.

50. *The Republic*, trans. Paul Shorey in *The Collected Dialogues of Plato*, ed. Edith Hamilton and Huntington Cairns (New York: Pantheon, 1961), 9.592b. Mr. Shorey emphasizes this point in his introduction, p. 576.

51. Benjamin Jowett's essay, "Plato's Philosophy," *The Works of Plato* (4 vols. in one: New York: Tudor, n.d.), p. li.

52. Paul Friedländer, *Plato I. An Introduction* (New York: Pantheon, 1958), p. 170. See chap. 8, "Dialogue," pp. 154–170, for a full discussion of the question.

53. Jowett, "Plato's Philosophy," p. lxi.

54. *Politics*, trans. Benjamin Jowett in *The Basic Works of Aristotle*, ed. Richard McKeon (New York: Random House, 1941), 2.1–8. Aristotle also criticizes the states of Phaleas of Chalcedon and Hippodamus, the 5th-century B.C. planner.

55. The best description of Zeno's *Republic* is found in *Diogenes Laertius. Lives of Eminent Philosophers*, trans. R. D. Hicks (LCL), II, 7.1, especially secs. 32–34. See also A. C. Pearson. *The Fragments of Zeno and Cleanthes* (London: C. J. Clay & Sons, 1891), especially the introduction and Fragments 149 and 162.

III : THE RENAISSANCE AND TRANSITION

1. Roger Ascham, *The Scholemaster*, in *English Works*, ed. W. A. Wright (Cambridge, England: Cambridge University Press, 1904), p. 213.

2. Benjamin Farrington, *Science in Antiquity* (London: Oxford University Press, 1950), chap. 9, "The Graeco-Roman World," pp. 200–220.

3. Charles Singer, *A Short History of Scientific Ideas to 1900* (Oxford: Clarendon Press, 1959), p. 132. For a brief but enlightening discussion of scientific development in the Roman civilization, see chap. 4, "The Failure of Inspiration. Science the Handmaid of Practice: Imperial Rome (50 B.C.–A.D. 400)," pp. 103–136.

4. *Seneca. Physical Science in the Time of Nero. Being a Translation of the Quaestiones Naturales of Seneca*, by John Clarke with notes by Sir Archibald Geikie (London: Macmillan, 1910). The

quotation is from the concluding chaps. 30 and 31 (pp. 304–306), after which Seneca deplores the progress of vice instead of truth, the degeneracy of society instead of the ascent to knowledge. It is a Stoic lament at the end.

5. Singer, *A Short History*, p. 108.

6. *A Translation of Dante's Eleven Letters*, with notes by Charles S. Latham (New York, 1892), Letter xi, "To Can Grande della Scala," secs. 15, 16.

7. For a comprehensive and detailed study of the period, see A. C. Crombie, *Medieval and Early Modern Science* (Cambridge, Massachusetts: Harvard University Press, 1961), vol. I, *Science in the Middle Ages: V–XIII Centuries*, vol. II, *Science in the Later Middle Ages and Early Modern Times: XIII–XVII Centuries.*

8. Sir Thomas More, *Utopia and a Dialogue of Comfort* (London: Dent, 1951), pp. 83–84.

9. See J. H. Hexter, *More's Utopia. The Biography of an Idea* (Princeton, New Jersey: Princeton University Press, 1952) for an analysis of More's attitudes. Hexter explains in detail More's realistic concern with economic problems and their relation to his ideals as a Christian humanist.

10. More, *Utopia*, p. 96.

11. R. P. Adams, "The Social Responsibilities of Science in *Utopia, New Atlantis*, and After," *Journal of the History of Ideas*, X (1949), 374–398.

12. Sir John Eliot, *The Monarchie of Man*, ed. Rev. Alexander B. Grosart (London, 1879), II, 134–135. Volume I includes biographical material and Grosart's interpretations of Eliot's work. Volume II contains Eliot's MS. which was found in the Harleian Collection and here printed for the first time.

13. Joseph Hall, *The Discovery of a New World* (*Mundus Alter et Idem*), ed. Huntington Brown (Cambridge, Massachusetts: Harvard University Press, 1937), pp. 87–88. This is the English translation by John Healey, c. 1609, which is the only complete translation of the work.

14. *Ibid.*, p. 100.

15. *Ibid.*, pp. 99–100. (Italics mine.)

16. J. Max Patrick. "Puritanism and Poetry: Samuel Gott," *University of Toronto Quarterly*, VIII (1939), 211–226. This article is primarily concerned with Gott's poetry and his discourse on poetic theory in vol. I, chaps. 3 and 4 of the *Nova Solyma.*

17. *Nova Solyma. The Ideal City or Jerusalem Regained*, introd., trans., and bibliography by Rev. Walter Begley (New York: Charles Scribner's Sons, 1902), I, chap. 1, bk. 2, pp. 168–169. Subsequent references will be to this edition and cited as *Nova Solyma.* The work, first published anonymously in 1648, was discovered by Begley, who attributed authorship to John Milton, but Stephen Jones later established Gott as the author. See Jones' article, "The Authorship of Nova Solyma," *The Library*, Ser. 3, I (1910), 225–238.

18. *Nova Solyma,* I, chap. 1, bk. 2, pp. 165–166. The printing press, compass, and gunpowder mentioned were the three favorite examples used to show the remarkable progress man had made beyond the ancients' knowledge, and Gott puts his finger on a sore point when he cites the inadequacy of knowledge regarding the compass. See Robert K. Merton, "Science, Technology and Society in Seventeenth-Century England," *Osiris,* IV (1938), 522–534, for a summary of research in magnetism, the compass, and various attempts to determine longitude.

19. François Rabelais, *The Histories of Gargantua and Pantagruel,* trans. J. H. Cohen (Baltimore, Maryland: Penguin, 1955), p. 157. The full description of the Abbey of Thélème is in *The Great Gargantua,* chaps. 52–58.

20. Erich Auerbach, *Mimesis. The Representation of Reality in Western Literature*, trans. Willard Trask (New York: Doubleday Anchor, 1957), p. 255.

21. *The Essays of Montaigne*, trans. E. J. Trechmann (2 vols. in 1; New York: Oxford University Press, 1946), I, bk. 1, p. 206. This passage, as translated into English by Florio (1603), was reproduced closely by Shakespeare in the *Tempest.* See Introduction.

22. Robert Burton, *The Anatomy of Melancholy* (London: Dent, 1948), I, preface, pp. 97–107. Subsequent references cited as *Anatomy* are from this text, based on the 1651–1652 edition, unless otherwise noted. This was the last edition to contain Burton's corrections, which generally meant addition rather than deletion of material.

23. *Ibid.*, II, 34–61.

24. Galileo Galilei, *Sidereus Nuncius* (Venice, 1610). There are two English translations of the work: *The Sidereal Messenger of Galileo Galilei,* trans. with introd. and notes by Edward Stafford Carlos (London, 1880), and *The Starry Messenger,* included in *Discoveries and Opinions of Galileo,* trans. with introd. and notes by Stillman Drake (New York: Doubleday Anchor, 1957).

25. Robert Burton, *Philosophaster,* trans. Paul Jordan-Smith (Stanford, California: Stanford University Press, 1931), II.v (p. 81). This is the first English translation. The first published edition was in 1862 from Burton's own Latin MS., printed privately by W. E. Buckley in an edition limited to 65 copies for a society of antiquarians (preface, Paul Jordan-Smith, pp. 11–12). Theoreticians mentioned by the character Philobiblos are identified (note 33, p. 232) as follows: Fracatorius (Hieronymus Fracastoro 1484–1553) was an astronomer, geologist, physicist, poet, and physician, a Veronese genius, whose *Homocentrica* (Venice, 1538) was a valuable contribution to astronomical science; Helisaeus (Eliseo) Roeslin was a German astronomer and scholar who lived at Frankfurt and, among other works, wrote *Theoria nova coelestium Meteoron* (Strassburg, 1578); Franciscus Patricius or Patrizi (1529–1597) lived at Caieta, Italy, and wrote several works on history, mathematics, and astronomy mentioned in *The Anatomy of Melancholy* ("Digression of Air"); Thaddeus Haggesius, according to a note in *The Anatomy,* wrote a book on *Metoposcopy* about 1578.

26. In the "Digression of Air" (*Anatomy,* II, 34–61) Burton listed the "new stars" of 1572, 1600, and 1604 and considered whether they were signs of "generation and corruption" in the world or merely orbs that have existed from the beginning and "show themselves at set times," and "have poles, axle-trees, circles of their own, and regular motions" as Helisaeus Roeslin contends. Other theories are also stated (pp. 50–51).

27. Anthony à Wood, *Athenae Oxoniensis . . . ,* ed. Philip Bliss (London, 1813), II, col. 652.

28. Burton, *Anatomy,* I, preface, p. 101.

29. Benjamin Farrington, *Francis Bacon, Philosopher of Industrial Science* (New York: Henry Schuman, 1949), p. 48.

30. Burton, *Anatomy,* I, 106. I have omitted Burton's Latin phrases from the quotation since they are not essential here to the meaning.

31. A Latin version of Ricci's commentaries had just been published by Trigault in 1615, and an extensive reader like Burton might well have encountered it. Burton's gloss is as follows: "Ad regendam rempub. soli literati admittuntur, nec ad eam rem gratia magistratus aut regis indigent, omnia explorata cujusque scientia et virtute pendent.—Riccius, lib. I, cap. 5" Paul Jordan-Smith, in his *Bibliographia Burtoniana* (Stanford, California: Stanford University Press, 1931), attributes this quotation to Matthew Ricci. Further ex-

amination of Ricci's work led me to the Trigault version entitled *De Christiana expeditione apud Sinas suscepta ab Societate Jesu,* which includes descriptions of the mores, laws, and institutions of 16th-century China. There is an interesting recent English translation of Trigault's work: *China in the Sixteenth Century, Journals of Matthew Ricci,* trans. from Latin by Louis J. Gallagher (New York: Random House, 1953).

32. J. Max Patrick, "Robert Burton's Utopianism," *Philological Quarterly,* XXVII, No. 4 (October 1948), 357–358. For a similar conclusion in a broader study of Burton's thought, see William R. Mueller, *The Anatomy of Robert Burton's England* (Berkeley, California: University of California Press, 1952).

33. *James Harrington's Oceana,* ed. S. B. Liljegren (Heidelberg: C. Winter, 1924). The text is based on a copy of the original edition in the Royal Library at Copenhagen; Liljegren consulted also a copy in the British Museum.

34. R. H. Tawney, "Harrington's Interpretation of His Age," The Raleigh Lecture on History, British Academy, read May 14, 1941. *Proceedings of the British Academy,* XXVII (1941), 205.

35. John Toland, *The Oceana of James Harrington, and His Other Works . . . with an Exact Account of his Life Prefix'd* (London, 1700), p. 18.

36. *Ibid.,* p. 161.

37. Because of the political upheaval in England during the 17th century, many idealistic proposals for correcting the disorder were made. Often these tracts are constitutional reforms and do not include the trimmings necessary for a utopia. Hobbes' *Leviathan,* and the less known Robert Filmer's *Patriarcha* are of this type. The anonymous *Free State of Noland* might have been included in the text, since it does have the fictional stage setting, but there is no concern with science either as a subject or as an approach to the utopia. For the interested reader, the available editions of these works are the following: Thomas Hobbes, *Leviathan* (London, 1651), reprinted (Oxford: Clarendon Press, 1952); *Patriarcha and Other Political Works of Sir Robert Filmer,* ed. Peter Laslett (Oxford: Basil Blackwell, 1949). The *Patriarcha* was not published until 1680, but Mr. Laslett shows on internal evidence that it was written during the period 1635–1642. *The Free State of Noland: or, The Frame and Constitution of the Happy, Noble, Powerful, and Glorious State. In which all Sorts and Degrees of People find their Condi-*

tion Better'd (London, 1701). With the exception of J. Max Patrick's article, "The Free State of Noland, A Neglected Utopia from the Age of Queen Anne," *Philological Quarterly*, XXV, No. 1 (January 1946), pp. 79–88, little attention has been paid to this work and there appear to be no definite clues to the identity of the author.

IV : THE NEW UTOPISTS

1. The phrase is Stephen Toulmin's in his essay "Seventeenth Century Science and the Arts," in a book of the same title, ed. Hedley Howell Rhys (Princeton, New Jersey: Princeton University Press, 1961), p. 7. This collection, resulting from the William J. Cooper lectures at Swarthmore College, includes interesting essays on the relation of science to literature, the visual arts, and musical thought of the century by Douglas Bush, James S. Ackerman, and Claude V. Palisca, respectively.

2. Charles Singer, *A Short History of Scientific Ideas to 1900* (Oxford: Clarendon Press, 1959), p. 221. See Francis R. Johnson, *Astronomical Thought in Renaissance England* (Baltimore, Maryland: Johns Hopkins Press, 1937), pp. 215–217. Gilbert immediately used his magnetic discoveries as a physical explanation of Copernicus' hypothesis of the earth's rotation (*De Magnete*, bk. 6), but he did not commit himself on the question of the earth's rotation around the sun. Johnson's study is informative for details concerning scientific writings in England from 1500 to 1645.

3. Singer, *A Short History*, p. 275. There is no definite evidence that Harvey knew Galileo, although it is held likely that, as an outstanding student at Padua, he would have visited the famous Galileo's classes. See Louis Chauvois, *William Harvey* (New York: Philosophical Library, 1957), pp. 68–69. Along with observation and experiment, however, Harvey did employ the quantitative method (without exactly accurate results) as one proof of the circulation of the blood; see his work *On the Motion of the Heart*, chap. 9.

4. Plutarch, *De Placitis philosophorum*, 3.3. This quotation is from Copernicus' dedication to Pope Paul III in his work *On the Revolutions of the Heavenly Spheres*, trans. Charles Glenn Wallis (Chicago, Illinois: Encyclopaedia Britannica, 1955), p. 508. This dedicatory letter by Copernicus should not be confused with the cautionary words to the reader, which the editor Andreas Osiander

prefixed to Copernicus' introduction. See George Sarton, *Six Wings. Men of Science in the Renaissance* (Bloomington, Indiana: Indiana University Press, 1957), pp. 59–60.

5. Singer, *A Short History*, pp. 27–28, 236–239.

6. *Ibid.*, p. 178.

7. Francis R. Johnson, in *Astronomical Thought*, lists and discusses popular works as well as more technical treatises of the 16th and first half of the 17th centuries. I cite Galileo and Harvey as outstanding examples.

8. For a discussion of this see Dorothy Stimson, *Scientists and Amateurs. A History of the Royal Society* (New York: Henry Schuman, 1948).

9. Professor Daniel J. Donno has prepared for this study the first complete English translation of the original 1602 Italian MS. Subsequent references will be cited simply as *City* and, unless otherwise noted, will be to his translation, which is based upon the edition of Norberto Bobbio (Turin: Einaudi, 1941) as reprinted by Romano Amerio in *Opere di Giordano Bruno e di Tommaso Campanella* (Milan: Ricciardi, 1956). Donno explains, however, that in the case of three brief, obscure, and not otherwise notable passages he chose to follow the punctuation supplied by Luigi Firpo in *Scritti Scelti di Giordano Bruno e di Tommaso Campanella* (Turin: Unione Tipographica-Editrice Torinese, 1949). Paragraphing and—so far as differences between Italian and English practice allow—capitalization are consistent with the text reprinted by Amerio. It is expected that this new translation will soon be published.

Until now there has been a variety of partial translations. William J. Gilstrap (who generally omitted astrological passages and a section on military matters) based his translation on the first printed edition of the Latin text: *Civitas Solis Poetica: Idea Reipublicae Philosophiae* (Frankfurt, 1623). This translation is carried by Glenn Negley and J. Max Patrick, *Quest for Utopia* (New York: Henry Schuman, 1952). The T. W. Halliday translation appeared in Henry Morley's *Ideal Commonwealths* (London, 1893) and later in Charles M. Andrews' *Ideal Empires and Republics* (New York: Aladdin, 1901). There is no indication of the Latin edition on which it is based, and the text is expurgated to fit Victorian standards of propriety. Mary Louise Berneri translated only parts of the 1602 Italian MS. for her *Journey Through Utopia* (Boston, Massachusetts: Beacon Press, 1951).

The Bobbio edition of 1941, selected by Professor Donno, had available ten of the eleven known MSS. of the work (the eleventh, in Vienna, was not available then), and Luigi Firpo's edition (1949) included three new MSS., but minimum changes in the text. The best earlier edition was Paladino's in 1920, based on collations of various MSS. edited to that date: Solmi's (1904), Kvacala's (1911), and Ciampoli's (1911). Miss Berneri comments (pp. 91–94) that a comparison of MSS. reveals that in the later Latin versions the *City* is more authoritarian and conforms more closely to ideas of the church; sexual freedom is more restricted, and astrology occupies a less important place. The definitive bibliography on Campanella's works is Luigi Firpo, *Bibliografia degli scritti di Tommaso Campanella* (Turin: Unione Tipographica-Editrice Torinese, 1940).

10. John Addington Symonds, *The Sonnets of Michael Angelo Buonarroti and Tommaso Campanella* (London, 1878), introd., p. 18.

11. Edmund G. Gardner, *Tommaso Campanella and His Poetry* (Oxford: Clarendon Press, 1923), p. 12. Although the dates are uncertain, Gardner says that "most of Campanella's poetry falls between the years 1600 and 1613." See Gardner, pp. 9–10, regarding sources of Campanella's poetry.

12. "Sonnet LVI," trans. Symonds, in *The Sonnets*, p. 174.

13. "Sonnet VI," trans. Symonds, p. 124.

14. See the discussion by Harald Höffding in *A History of Modern Philosophy*, trans. B. E. Meyer (New York: Humanities Press, 1950), I, 94.

15. Lynn Thorndike, *A History of Magic and Experimental Science* (New York: Columbia University Press, 1941), VI, 370.

16. Gardner, *Tommaso Campanella*, p. 22. Cf. Marjorie H. Nicolson, *Science and Imagination* (Ithaca, New York: Cornell University Press, 1956), pp. 24–25. The letter from Campanella was dated January 13, 1611, and quoted in *Le Opere*, XI, 21–26.

17. *The Defense of Galileo of Thomas Campanella*, trans., introd., and ed. Grant McColley in *Smith College Studies in History*, XXII (April–July 1937), introd., p. 12. Subsequent references will be cited as *The Defense*.

18. *City*, p. 55. This passage was not changed in later editions of the work. Edmondo Solmi (1904) based his edition on an examination of six MSS., and showed the chief variants between the Italian and two Latin versions in his notes. At this point in the text he indicates no variants. Later editors, Firpo and Amerio, are silent on

this point, too. Dr. Donno has found no evidence to the contrary.

19. Singer, *A Short History*, p. 212. Cf. McColley's introd. (p. 31) to *The Defense*, in which he says that Copernicus made 27 observations in order to compute better the orbits of the planets. This is a relatively small number of observations compared to those of Brahe and others. See also J. L. E. Dreyer, *A History of Astronomy from Thales to Kepler*, rev. with foreword by W. H. Stahl (New York: Dover, 1953), chap. 13, "Copernicus," pp. 305–344.

20. Singer, *A Short History*, p. 91.

21. See note 16 in this Chapter.

22. *The Defense*, introd., p. 23. The passages in *The Defense* are found on pp. 30, 33, 35, 50, 50, 52.

23. *Ibid.*, p. 54.

24. See Professor Patrick's elaboration of this in Negley and Patrick, *Quest for Utopia*, pp. 314–315.

25. *City*, p. 25. For the general background of Campanella's astrological thought, see Don Cameron Allen, *The Star-Crossed Renaissance* (Durham, North Carolina: Duke University Press, 1941).

26. Thorndike, *History of Magic*, VI, 172–177, has an excellent description of Campanella's astrological works.

27. Ernest Hatch Wilkins, *A History of Italian Literature* (Cambridge, Massachusetts: Harvard University Press, 1954), pp. 299–300. Cf. Gardner, *Tommaso Campanella*, especially pp. 5–9.

28. Thorndike, *History of Magic*, VI, 173. Campanella also believed that the grand celestial conjunction of 1603 was to produce a universal kingdom of saints (Negley and Patrick, *Quest for Utopia*, p. 314).

29. "Sonnet III," trans. Symonds, in *The Sonnets*, p. 121.

30. Thorndike, *History of Magic*, VI, 371.

31. Gardner, *Tommaso Campanella*, p. 6.

32. Thorndike, *History of Magic*, III, 6.

33. *Ibid.*, VI, 174.

34. *Ibid.*, p. 173. The quotation is from a letter of 1606 to Pope Paul V.

35. *City*, p. 66.

36. Singer, *A Short History*, p. 237. A good discussion of astrology will also be found in Sarton, *Six Wings*, pp. 72–76.

37. Singer, *A Short History*, p. 216.

38. Lynn Thorndike, *The Place of Magic in the Intellectual History of Europe* (New York: privately printed, 1905), p. 22.

39. Mary Louise Berneri in *Journey* points this out and Charles Andrews in *Ideal Empires* makes the same statement. It is subject to argument only on the question of whether one takes the date of composition (MS. date) or the actual date of publication; in the latter case, Andreae's *Christianopolis*, first printed in 1619, would precede Campanella's *City of the Sun*, first published in 1623 in Latin. Since Campanella's work circulated for many years in various MS. forms, from 1602 to 1623, I have placed him first chronologically.

40. *Johann Valentin Andreae's Christianopolis*, trans. and introd. Felix E. Held (diss., University of Illinois, privately printed, 1914), chap. 11, "Metals and Minerals," pp. 154–155. This is the only English translation of the original *Reipublicae Christianopolitanae Descriptio* (Strassburg, 1619), which will be found in the Royal Library, Copenhagen. J. F. C. Richards has compared the translation with the original and judged Held's work most adequate. Hereafter references to the text will be to this translation and cited simply as *Christianopolis;* references to Held's introduction will be cited as Held, introd. The text is composed of 100 short sections or chapters.

41. *Christianopolis*, chap. 44, "The Laboratory," p. 196.

42. See Ferdinand Maack, *Chymische Hochzeit: Christiani Rosencreutz Anno 1459* (Berlin: Herman Barsford, 1913), introduction, for this and a full discussion of the movement.

43. *Ibid.*, introd., pp. 36–38. Cf. Walter Weber, "Nachwort," in *Johann Valentin Andreae, Die Chymische Hochzeit des Christian Rosenkreuz Anno 1459* (Stuttgart: Freies Geistesleben, 1957), p. 181, for MSS. dates and editions.

44. Maack, *Chymische Hochzeit*, introd., p. 39.

45. Weber, "Nachwort," p. 182.

46. Maack, *Chymische Hochzeit*, introd., p. 43.

47. *Ibid.*, note, p. 44.

48. *Ibid.*, introd., p. 43. Andreae also wrote a letter to Comenius in 1629 concerning this society.

49. *Ibid.*, p. 36, Cf. Weber, "Nachwort," p. 181.

50. Singer, *A Short History*, p. 13.

51. Maack, *Chymische Hochzeit*, introd., p. 40. The statement is based on Maack's quotation from Karl Kiesewetter, an authority on Rosicrucianism.

52. Although there were, according to Maack (introd., p. 38), two English translations of the two works (*Fama Fraternitatis* and

Confessio Fraternitatis), in 1652 and 1887, I have not been able to locate them. References are here translated from the best German texts available: the *Fama*, trans. Walter Weber, in *Johann Valentin Andreae* . . . (Stuttgart: Freies Geistesleben, 1957), pp. 103–118, and the *Confessio*, trans. Ferdinand Maack, in *Chymische Hochzeit* . . . (Berlin: Herman Barsford, 1913), pp. 67–84. Subsequent references will be cited simply *Fama* and *Confessio*.

53. Held, introd., pp. 39, 54, 59, 72, 119, 120, 121. In trying to prove Andreae's influence on Bacon's *New Atlantis*, Held cites, for example, a passage about the Brothers' traveling abroad to collect data (p. 73). He implies that it is scientific data, but such is not actually stated in the *Fama*. He also translated (p. 11) the fraternity as a "college" while "cloister" or "convent" seems closer to Andreae's intent.

54. *Fama*, p. 103.

55. *Ibid.*, p. 105.

56. *Ibid.*, p. 107.

57. According to the *Confessio*, the Father was born in 1378 and died when he was 106, which made the year 1484. He was to have been revealed in 120 years, so the opening of the grave occurred in 1604, the year in which the *Chymische Hochzeit* was supposedly written which, according to Weber, described the last initiation of the Father. Maack (introd., p. 41) gives 1602 as the year the *Hochzeit* was written.

58. The *Fama* circulated widely, appearing rapidly in several languages: Latin, 1614; Dutch, 1615; French, 1616; Italian, 1617; English, 1652 (Maack, introd., pp. 37–38, and Held, introd., p. 11). The same exhortation, promising that no one was to be excluded "from the blessedness of the Brotherhood's principles which should free the whole world," occurs in the *Confessio*, pp. 74–75.

59. *Confessio*, p. 82.

60. Maack (introd., p. 18) says the movement started in Germany under the physician and alchemist Michael Maier (1570–1622); according to Rudolf Steiner, its spiritual roots date back to the 4th century, and its appearance physically to the year 1050 (Weber, "Nachwort," p. 180).

61. Carl Grüneisen, *Die Christenburg von Johann Valentin Andreae*, in *Zeitschrift für die historische Theologie* (Leipzig, 1836), VI, introd., p. 246. Here Grüneisen summarizes Andreae's purpose as indisputably "to contrast the true secret, the basis and kernel of genu-

ine science, and the deep spirit of the wonders of the realm of nature . . . with vain secrets and valueless brooding, quibbling, and trifling with nature." Held cites this (introd., note, p. 121) as evidence of Andreae's support of real science, but he fails to observe that Grüneisen was speaking of Rosicrucians as countering the false ideas.

62. *Christianopolis*, preface, pp. 137–138. I have not corrected the Held translation in sentence structure; Andreae's Renaissance Latin in the original text frequently includes run-on sentences.

63. Wilhelm Gussmann, *Reipublicae Christianopolitanae Descriptio*, in *Zeitschrift für kirchliche Wissenschaft und kirchliches Leben* (Leipzig, 1886), VII, 442.

64. Grüneisen, *Die Christenburg*, pp. 254ff. The epic is composed in iambic feet, couplet rhyme, and is generally tetrameter, idiomatic German. The whole is composed of 40 songs, 50 lines in each.

65. In this opinion I differ from Professors Negley and Patrick (*Quest for Utopia*, p. 290), who consider Andreae's allegorical content greater than that in Campanella's *City*, and from Andreas Voight, who reads the *Christianopolis* as largely allegorical. In support of my position, see Held, introd., pp. 25–27.

66. It may have been Hafenreffer, a mutual friend of Kepler's and Andreae's, who brought them together. This suggestion is Christoph Sigwart's, who discusses Campanella's sonnets, translated by Andreae, in *Kleine Schriften* (Freiburg, 1889), I, 208.

67. Carl Hüllemann, *Valentin Andreae als Paedagog* (Leipzig, 1884), p. 19. He quotes from Andreae's *Collectaneorum Mathematicorum Decades XI*.

68. *Ibid.*, p. 19.

69. Sarton, *Six Wings*, p. 64.

70. A. R. Hall, *The Scientific Revolution 1500–1800. The Formation of the Modern Scientific Attitude* (Boston, Massachusetts: Beacon Press, 1956), p. 65.

71. Sarton, *Six Wings*, p. 64.

72. Hüllemann, *Valentin Andreae*, pp. 20–21.

73. *Ibid.*, p. 14.

74. *Ibid.*, pp. 4–5. Hüllemann takes his facts from Andreae's Latin autobiography edited by F. H. Rheinwald under the title *Joannis Valentini Andreae Theologi Württenbergensis vita, ab ipso conscripta* (Berlin, 1849), pp. 220, 229.

75. Various persons have been suggested as the go-between for Andreae and Adami. Gussmann thinks "the most likely" person was Andreae's best friend, W. von der Wense, but "it could also be Besold" (p. 438). Christoph Besold started a translation of Campanella's *Spanish Monarchy* at the same time Andreae translated the sonnets. And Besold also gave Andreae the use of his fine library, so he may be the more likely person to have introduced Andreae to Tobias Adami in 1618, as Sigwart claims (p. 174).

76. Andreae's translation appeared in *Geistliche Kurzweil* (Strassburg, 1619), the same year the *Mythologia* was issued.

77. Held, introd., p. 21.

78. The question of influence among the three utopias of Andreae, Campanella, and Bacon has engaged many scholars to the point of no conclusion. I think the utopias are each highly individualistic; some similarities may be explained as evidence of the general challenge of contemporary thought. For the position that Andreae "follows slavishly" Campanella's *City*, see Robert Mohl, *Die Geschichte und Literatur der Staatswissenschaften* (Graz, 1855), reprinted (Academische Druck- und Verlagsanstalt, 1960), I, 184ff. Christoph Sigwart agrees with Mohl. Gussmann has considered their arguments and refutes them ably. Felix Held, in the introduction to *Christianopolis*, tries to show Andreae's influence on both Bacon and Campanella, but the evidence is highly hypothetical. In the same way Eleanor Blodgett, in an article, "Bacon's New Atlantis and Campanella's Civitas: A Study in Relationships," *PMLA*, XLVI (1931), 763–780, attempts unconvincingly to show Bacon's adoption of Campanella's ideas. Often these critics are not well grounded in the utopian genre—what is typical of utopias that may stem from a variety of sources—and so they find parallels between authors in what is merely part of the utopian pattern. In the cases considered here there is simply insufficient evidence to prove direct influence. Only Andreae we know *could* easily have seen Campanella's utopia, and Bacon mentions Campanella's name in a list of natural philosophers.

79. *Historia Naturalis et Experimentalis*, in *The Works of Francis Bacon*, ed. Spedding, Ellis, and Heath (London, 1887), II, 13–14. The Latin quotation which begins the work is translated by R. F. Jones in *Francis Bacon* (New York: Odyssey Press, 1937), p. 368. Ellis stated that this was the only reference to Campanella in all

Bacon's works. I have found no other. For a detailed analysis of Bacon's attitude toward the various philosophers named, see Fulton H. Anderson, *The Philosophy of Francis Bacon* (Chicago, Illinois: University of Chicago Press, 1948). This is an excellent, comprehensive study of Bacon's philosophic writings, and includes Bacon's references to Telesio.

80. *Novum Organum,* in *The Works of Francis Bacon,* ed. Spedding, Ellis, and Heath (London, 1887), IV, 1.116. Future references to this and Bacon's other works are to this edition.

81. *Ibid.,* 1.54.

82. Singer, *A Short History,* p. 222.

83. *The Works of Francis Bacon,* II, 5.

84. *Novum Organum,* 1.64.

85. *Ibid.,* 2.39.

86. Singer, *A Short History,* p. 221.

87. John William Adamson, *Pioneers of Modern Education* (Cambridge, England: Cambridge University Press, 1921), p. 9.

88. *The Works of Francis Bacon,* XIV, 131. The letter was dated October, 1620, so Bacon acted quickly after the publication of the *Novum Organum* (1620). Spedding does not mention further correspondence concerning this from Wotton.

89. *Ibid.,* pp. 35–36. Spedding observes that two other letters from Matthew about Galileo are listed in Stephens' catalog but not extant.

90. *Ibid.,* p. 35. Spedding attributes this idea to Ellis, but neither editor apparently realized that Galileo had just reissued the work. For this, see *Discoveries and Opinions of Galileo,* trans. with introd. and notes by Stillman Drake (New York: Doubleday Anchor, 1957), pp. 220–221.

91. *The Advancement of Learning,* in *The Works of Francis Bacon,* III, 1, p. 289.

92. See Don Cameron Allen, *The Star-Crossed Renaissance,* pp. 149–153.

93. *Novum Organum,* 1.65. See Anderson, *The Philosophy of Francis Bacon,* pp. 124–131, for a careful analysis of Bacon's attitudes toward Plato.

94. *Novum Organum,* 1.60.

95. R. F. Jones, *Francis Bacon,* introd., p. 23. Bacon explains his animistic theory in the *Historia Vitae et Mortis* (p. 94) and in the

History of Density and Rarity (p. 237) in *The Works of Francis Bacon*, II.

96. William Rawley, "To the Reader," prefixed to *Sylva Sylvarum*, in *The Works of Francis Bacon*, II, 336.

97. Mary Louise Berneri, in *Journey*, p. 94, makes this point.

98. Held, introd., p. 13. At Nuremberg with three friends in 1628 Andreae again tried to establish a Christian Union, but this one did not enjoy the long life of his earlier venture. See G. H. Turnbull, *Hartlib, Dury, and Comenius. Gleanings from Hartlib's Papers* (London: Hodder and Stoughton, 1947), p. 73.

99. This attempt of Bacon's is discussed in James Spedding, *The Letters and The Life of Francis Bacon* (London, 1889), pp. 25, 66.

100. The words were written about Francis Bacon by James Spedding, preface to the *New Atlantis*, in *The Works of Francis Bacon*, III, 122.

V : THE DOCTRINE SPREADS

1. Joseph Glanvill, "Anti-fanatical Religion, and Free Philosophy. In a Continuation of the New Atlantis," *Essays on Several Important Subjects in Philosophy and Religion* (London, 1676), p. 7. This essay will be cited hereafter as Glanvill, "Continuation."

2. Marie Louise Berneri, in her *Journey Through Utopia* (Boston, Massachusetts: Beacon Press, 1951), pp. 321–322, cites several odd publications using the word "utopia" in their titles: *The Utopian*, a magazine of precarious existence; *A Charge from Utopia*, a religious tract; *The Six Days Adventure or the New Utopia*, purporting to describe an attempt at a commonwealth of women and enacted at the Duke of York's Theatre in 1671; a curious political pamphlet published in 1647 as a letter from the "King of Utopia to the citizens of Cosmopolis, the metropolitan city of Utopia"; and *Canary Birds Naturaliz'd in Utopia*, a satirical canto printed about 1709.

3. Dorothy Stimson, *Scientists and Amateurs: A History of the Royal Society* (New York: Henry Schuman, 1948), p. 67, mentions this as evidence of the attitude toward the Royal Society and its activities, which were considered both "futile in themselves and also dangerous to the state and to the established church." Earlier in his life, in 1658 during travels abroad, Oldenburg had written Hartlib and cautioned him to show his letters to no one and burn them after

reading. G. H. Turnbull suggests that this may have been because his letters quite often contained political news. See his article, "Samuel Hartlib's Influence on the Early History of the Royal Society," *Notes and Records of the Royal Society*, X (1953), 116.

Because of the Order of 1655 against scandalous books and pamphlets, James Harrington had great difficulty in obtaining permission to print his *Oceana*, according to H. F. Russell Smith in *Harrington and His Oceana* (Cambridge, England: Cambridge University Press, 1914), p. 9. S. B. Liljegren, in his introduction to *James Harrington's Oceana* (Heidelberg: C. Winter, 1924), contests Smith's assertion, but the phrase remains in Harrington's "Epistle to the Reader," in which he refers to the "spaniel questing" which had interfered with printing his book.

4. Hartlib's papers are the material used by G. H. Turnbull in *Hartlib, Dury, and Comenius. Gleanings from Hartlib's Papers* (London: Hodder and Stoughton, 1947). Turnbull explains in the preface that after Hartlib's death (which he dates 1662, though there is question concerning this) his papers were purchased by a friend, William Brereton. They were found in 1667, at Brereton's estate in Cheshire, by John Worthington, who arranged them carefully in separate bundles. The recently found box contains such bundles, which Turnbull believes to be those made by Worthington. Apparently the new materials are under the care of Lord Delamere, whom Turnbull thanks for permission to use them. He gives no further details as to how or where the discovery was made. The papers are especially valuable in adding information about Cyprian Kinner, Dury, and Comenius as well as Hartlib. And there are other important materials deserving to be published, such as Hartlib's correspondence with Henry More (1648–1655), which was assumed to be nonexistent.

Hartlib's publications and manuscripts are often difficult to identify because of similarity of theme and the frequent practice of not signing such tracts. The papers found also include many works by his friends with parallel interests. For lists of Hartlib's own publications, separated from those he caused to be printed or received from others, see Turnbull, *Hartlib, Dury, and Comenius*, and H. Dircks, *A Biographical Memoir of Samuel Hartlib . . .* (London, 1865). Turnbull is, of course, more up to date and includes the papers recently found. Dircks mentions another plan of Hartlib's for a philosophical college to be called "Antlantis" which Hartlib included in

A Memorial for the Advancement of Learning. I have not included this in the present study because Turnbull attributes the plan to Dury, and I have not been able to determine authorship.

5. In the *Harleian Miscellany* (London, 1810), VI, 27–36. This work is unsigned and undated. It is typical of the period and lists items requiring financial support for experimentation, including a seed that affords three crops a year, a way to make grapes ripen as early as cherries, and treatments for gout, scurvy, and "the stone." All promise practical success. Hartlib is listed as author of a *Cornu Copia* (London, 1652) in the British Museum Catalogue, but the copy has no title page. Turnbull thinks some parts correspond to passages in a document by Adolphus Speede, and, because the remainder cannot be identified, he does not attribute the work to Hartlib. Suffice it to say that the cornucopia was frequently used as an emblem by utopists and other writers of the period. It appears again in the continuation of the *New Atlantis* by R. H., who ends a discourse between two gentlemen in a shady grotto by the intrusion of two gardeners who enter symbolically with "a large cornucopia in each hand full of all the variety of fruit the season then afforded."

6. W. H. G. Armytage, *Civic Universities: Aspects of a British Tradition* (London: Ernest Benn, 1955), p. 101. The statement was made by John Dury, whose efforts to reform education and to unite Protestantism throughout Europe were encouraged by Hartlib.

7. A. R. Hall, *The Scientific Revolution 1500–1800: The Formation of the Modern Scientific Attitude* (Boston, Massachusetts: Beacon Press, 1956), p. 193.

8. Dorothy Stimson, "Comenius and the Invisible College," *Isis*, XXIII (1935), 373. Miss Stimson's main thesis is that Comenius was not the major instigator of the early meetings, as several critics have assumed. Instead she cites (note, p. 376) John Wallis' account of the early activities leading to the Royal Society's establishment, in which he credits Haak with the leadership. This Wallis stated in some passages concerning his own life, January 29, 1696–1697, in T. Hearne, *Works* (London, 1810), III, clxii.

9. Comenius, in *The Labyrinth of the World and the Paradise of the Heart,* chap. 10, sec. 11, also names Campanella and Ramus in debate with the peripatetics before the queen, Wisdom. This and other interesting works of Comenius have recently been made easily available in a UNESCO publication: *John Amos Comenius: Selec-*

tions, introd. Jean Piaget (Switzerland: Centrale Lausanne S.A., 1957).

10. Wilhelm Gussmann, *Reipublicae Christianopolitanae Descriptio,* in *Zeitschrift für kirchliche Wissenschaft und kirchliches Leben* (Leipzig, 1886), VII, 471. Comenius also speaks of Andreae with great respect in his *Novissima linguarum methodus* (chap. 29). Gussmann further suggests direct influence from Andreae's *Christianopolis* and *Theophilus* (concerning education) on Comenius' thought. He especially mentions evidence of the *Christianopolis* in Comenius' *Labyrinth of the World and Paradise of the Heart,* which was an allegorical search through the ills of the world on the order of Bunyan's *Pilgrim's Progress.*

11. Gussmann, *Reipublicae,* p. 442.

12. Boyle's letter, dated April 8, 1647, is found together with several others from Boyle to Hartlib in Thomas Birch's "Life of the Honourable Robert Boyle" prefixed to *The Works of Robert Boyle,* ed. Thomas Birch (London, 1744), I, 22–23. In an undated letter (p. 28) Boyle also inquires about Hartlib's progress toward the success of his "college" and offers to contribute "what thoughts and observations of mine I shall judge useful in reference to so glorious a design." Boyle's esteem for "honest Hartlib," as he twice called him, was further evident by the dedication of Title XIII in his *General History of the Air* (V, 124) to Hartlib. For Hartlib's letters to Boyle during the years 1647–1659 see *The Works,* VI, 76–136. These and other letters collected in this edition show clearly the web of communications among Glanvill, Petty, John Wilkins, Oldenburg, Evelyn, and many others working to advance science and the Royal Society.

13. R. W. Gibson and J. Max Patrick, "Bibliography of Utopiana," in *St. Thomas More: A Preliminary Bibliography,* compiled by R. W. Gibson (New Haven, Connecticut: Yale University Press, 1961), note 650, p. 329.

14. John William Adamson, *Pioneers of Modern Education* (Cambridge, England: Cambridge University Press, 1921), p. 49. The quotation is from *J. A. Comenii Opera Didactica Omnia* (Amsterdam, 1657), 1.432. See Adamson, chap. 3, "Bacon and Comenius" (pp. 46–57), for a more complete discussion.

15. George Sarton, *Six Wings: Men of Science in the Renaissance* (Bloomington, Indiana: Indiana University Press, 1961), p. 60.

16. Stimson, "Comenius and the Invisible College," p. 383.

17. Adamson, *Pioneers,* pp. 50–51. Both analogies are from the *Didactica Magna.*

18. Robert Fitzgibbon Young, *Comenius in England* (London: Oxford University Press, 1932), pp. 54–55. This is an interesting study including several relatively unusual documents of Comenius (one translated into English for the first time) and his plans for the higher education of Indians in New England and Virginia.

19. *Ibid.,* p. 50.

20. Samuel Hartlib, *A Description of the famous Kingdom of Macaria . . .* , in *Harleian Miscellany* (London, 1808), I, 580–585. The author's name is not given, but the text is from the original quarto, containing 15 pages (London, 1641). Macaria, a word of Greek derivation, means "happiness or bliss." Subsequent references will be to this edition and cited as *Macaria.*

21. In Hartlib's papers are two MS. copies each of Andreae's *Christianae Societatis Imago* and *Christiani Amoris Dextera Porrecta* which were published in 1620 and had been thought by Kvacala, a critic on Andreae, to be no longer extant. Turnbull in *Hartlib, Dury, and Comenius,* note, p. 74, assumes that these MSS. were the ones translated into English in 1647 and 1657 by I. Hall, late of Grayes Inn. The first 1647 edition was dedicated to Mr. S. Hartlib. Robert Boyle mentions in his letter of March 19, 1646–47 (*The Works,* I, 22) that he has received these two tracts from Hartlib, and in the letter dated April 8, 1647, he says that he has read the *Imago Societatis* with pleasure, believes it to be of public value, and is delighted that Mr. Hall is so engaged in the endeavor.

22. Stimson, "Comenius and the Invisible College," p. 379.

23. *Macaria,* p. 583.

24. Turnbull, *Hartlib, Dury, and Comenius,* p. 73.

25. *The Advice of W.P. to Mr. Samuel Hartlib, for the Advancement of some particular Parts of Learning,* in *Harleian Miscellany* (London, 1810), VI, 1–14. The *Advice* was first printed in 1648 as a quarto containing 34 pages.

26. Abraham Wolf, *A History of Science, Technology and Philosophy in the 16th and 17th Centuries* (London: George Allen and Unwin, 1935), pp. 587, 598.

27. *Ibid.,* p. 598. The quotation is from the preface of Petty's *Political Arithmetick,* which was finished in 1676 but not published until 1690. Other studies of this type that he published were *Observations upon the Dublin Bills of Mortality* (1681) and *Essays in*

Political Arithmetick (1683–1687). Petty was one of three men in the period working in this field. John Graunt and Gregory King preceded and followed, respectively.

28. See the various works of Marjorie Hope Nicolson, including *Newton Demands the Muse* (Princeton, New Jersey: Princeton University Press, 1946), *Voyages to the Moon* (New York: Macmillan, 1948), and *The Breaking of the Circle* (Evanston, Illinois: Northwestern University Press, 1950). The preface to her articles collected in *Science and Imagination* (Ithaca, New York: Cornell University Press, 1956) is helpful in mentioning other authors writing on science and literature.

29. The best and most recent treatment of Cowley's thought is Robert B. Hinman, *Abraham Cowley's World of Order* (Cambridge, Massachusetts: Harvard University Press, 1960). This study discusses thoroughly the scientific developments of the century as they were understood and interpreted by Cowley. Hinman shows the unified approach of Cowley and his use of all materials and new knowledge in his poetry, and attempts to answer critics like Douglas Bush and Basil Willey, who have suggested that Cowley's scientific interest tended to sterilize his poetic metaphor.

30. "Of Agriculture," in *Abraham Cowley: Essays, Plays, and Sundry Verses*, ed. A. R. Waller (Cambridge, England: Cambridge University Press, 1906), p. 405.

31. Canto 6.66 in *The Works of Sir William Davenant* (London, 1672). *Gondibert,* which was never completed, was first published in 1651 and reissued in 1653. Astragon's House is described in bk. 1., cantos 5–6.

32. Cowley's notes state that Mahol is taken from I Kings 4.13. (This verse says that Solomon's wisdom is greater even than that of the Sons of Mahol.) Nathan and Gad were prophets in the time of David. Cowley adds that their particular professorships were "a voluntary gift of mine to them."

33. Abraham Cowley, *Davideis,* in *Poems,* ed. A. R. Waller (Cambridge, England: Cambridge University Press, 1905), bk. 1, p. 260. This edition contains the text of the first collected edition of 1668, which Thomas Sprat edited. At the end Waller shows variations noted in collation of this text with the folio of 1656, the volume of 1663, and others.

34. Abraham Cowley, *A Proposition for the Advancement of Experimental Philosophy*, in *Complete Works in Verse and Prose of*

Abraham Cowley, ed. Rev. Alexander B. Grosart (Edinburgh, 1881), II, 288. This reproduces the original text and preface published in London, 1661.

35. Joseph Glanvill, *Scepsis Scientifica* (1665), ed. John Owen (London, 1885), p. 65.

36. *History of the Royal Society*, ed. Jackson I. Cope and Harold Whitmore Jones (Saint Louis, Missouri: Washington University Studies, 1959), p. 35. The text here is from the original 1667 edition, which had not been reprinted in its entirety since 1734.

37. The influence of Bacon on the Royal Society has been thoroughly discussed by many scholars; the majority give him considerable credit for the Society, which thought of itself as carrying out the plans of the great Lord Verulam. However, Stephen Penrose, Jr., in *The Reputation and Influence of Francis Bacon in the 17th Century* (diss., Columbia University, privately printed, 1934), is an exception; he feels that Bacon's role has been overemphasized and that the Society owes its origin rather to the academy pattern developing on the Continent.

38. The identity of the author has not been determined. Dr. J. Max Patrick says that Robert Hunt has been suggested, but this is most unlikely unless Hunt wrote the tract when very young. I think it possible that Richard Haines (Haynes) wrote it. His dates (1633–1685) are acceptable, and he wrote several works of a utopian nature, such as a model of government for the good of the poor and the wealth of the nation, proposals for restoration of the woolen industry, and the improvement of workhouses, almshouses, and hospitals. He also signed himself "R. H." on the last proposal. In addition, Richard Haynes received patent rights on February 3, 1672, for "a way to sever, divide, & make cleane the seed called nonsuch trefoyle or hopclover. . . ." In the utopian work included in this study, the author reveals great interest in and knowledge of agricultural crops and herbs. The evidence is not conclusive but is worthy of further examination.

39. Joseph Glanvill, "Continuation." Glanvill also wrote another continuation entitled "Bensalem: Being a Description of a Catholic and free Spirit both in Religion and Learning, in a continuation of the Story of Lord Bacon's 'New Atlantis.'" This is an unpublished MS., n.d., now in the University of Chicago library. It is considered to be earlier than the continuation of 1676 and is dated "1675" for the library collection. The tract is especially concerned with reli-

gious sects and the "enthusiasts" Glanvill so thoroughly disliked. It seemed best to include in this study the published work of 1676, which is more general and covers several aspects of society.

40. Robert K. Merton, "Science, Technology and Society in Seventeenth Century England," *Osiris,* IV, 384–385. Cf. A. R. J. P. Ubbelohde's essay, "The Beginnings of the Change from Craft Mystery to Science as a Basis for Technology," *A History of Technology,* ed. Charles Singer *et al.* (Oxford: Clarendon Press, 1958), IV, 663–681. Ubbelohde comments (p. 665) on the growth of interest in natural science on the part of active and socially prominent leaders in the community, which was an essential factor in the introduction of scientific procedures into modern technology and most important, of course, to the general development of science itself.

Thomas Sprat, in his *History of the Royal Society,* ed. Cope and Jones, pp. 151–152, notes the increasing interest in science during the reign of Charles II and compares this period with that under James I when men "in imitation of the *King* . . . chiefly regarded the matters of *Religion,* and *Disputation:* so that even my Lord *Bacon,* with all his authority in the *State,* could never raise any *Colledge of Salomon,* but in a *Romance.*" He traces the developing interest until his day (1667) and says that now "there is a universal *desire,* and *appetite* after *knowledge.*"

VI : UTOPIAN ANSWERS TO AUTHORITY

1. *The Works of Gerrard Winstanley,* ed. with introd. by George H. Sabine (Ithaca, New York: Cornell University Press, 1941), pp. 501–600. The text is complete and taken from the original (London, 1652). Quotations are from this edition and subsequent references will be listed as Winstanley.

2. For an informative discussion on Israel as the model, see Hans Kohn, *The Idea of Nationalism. A Study in its Origins and Background* (New York: Macmillan, 1944), chap. 4, "Renaissance and Reformation."

3. *Johann Valentin Andreae's Christianopolis,* trans. and introd. Felix Held (diss., University of Illinois, privately printed, 1914), chap. 47, "The Natural Science Laboratory," p. 201. (Italics mine.) This work is cited as *Christianopolis.*

4. Basil Willey, in *The Seventeenth Century Background* (New York: Doubleday Anchor, 1953), has an excellent chapter on Joseph

Glanvill (pp. 174–205) in which he cites this quotation from *The Vanity of Dogmatizing* (p. 172).

5. Petty describes in lively fashion the boys' and girls' interests and imitative behavior, including the little female's propensity to act like a mother with her gossips, in his *Advice of W. P. To Mr. Samuel Hartlib, for the Advancement of some particular Parts of Learning*, in *Harleian Miscellany* (London, 1810), VI, 13.

6. *An Humble Motion*, p. 27. The statement is quoted by R. F. Young, *Comenius in England* (London: Oxford University Press, 1932), introd., p. 1.

7. John Wallis, "Letter to Rev. Thomas Smith, Dr. in Divinity, late Fellow in Magdalen College, Oxford. Oxford, Jan. 29, 1696/7," in Thomas Hearne, *Peter Langtoft's Chronicle* (Oxford, 1725), I, Publisher's Annex to His Preface, pp. 147–150. Phyllis Allen cites this in "Scientific Studies in Seventeenth Century English Universities," *Journal of the History of Ideas,* X (1949), 228–229.

8. Lynn Thorndike, *A History of Magic and Experimental Science* (New York: Columbia University Press, 1941), VI, chap. 31, "Post-Copernican Astronomy," pp. 3–66. Thorndike describes various problems in the acceptance of the new theory and feels that the "dead weight of pedagogical tradition and inertia did far more to delay the spread and general acceptance of the Copernician hypothesis than any religious opposition to it" (p. 7).

9. See Phyllis Allen, "Scientific Studies," pp. 219–253, for facts concerning the universities' adjustment to scientific studies. Also see Francis R. Johnson, *Astronomical Thought in Renaissance England* (Baltimore, Maryland: Johns Hopkins Press, 1937). Although Johnson gives evidence of more scientific activity at the universities than my brief summary implies, he agrees that it was mainly an indirect influence, that London was the center for such activities, and that it was the sporadic influence of individuals at the universities who attracted disciples to their mathematical and astronomical theories. He therefore concludes for his period to 1645 that "the majority, even among university students, acquired their detailed knowledge of the principles of astronomy, such as it was, wholly outside the regular university curriculum" (p. 12).

10. Allen, "Scientific Studies," p. 247.

11. Godfrey Davies, *The Early Stuarts 1603–1660* (Oxford: Clarendon Press, 1952), pp. 351–352. The Laudian Statutes of 1636 remained in force for more than two centuries at Oxford, and the

Statutes of 1570 continued to govern Cambridge on similar oligarchical lines. For detailed studies of the two universities, see Charles E. Mallet, *A History of the University of Oxford* (London: Methuen, 1924–1927), II, "The Sixteenth and Seventeenth Centuries," and James Bass Mullinger, *The University of Cambridge* (Cambridge, England: Cambridge University Press, 1873–1911). The last part of vol. II and vol. III cover the period of this study. For a more recent treatment, see William T. Costello, S.J., *The Scholastic Curriculum at Early Seventeenth-Century Cambridge* (Cambridge, Massachusetts: Harvard University Press, 1958), especially chap. 3, "The Undergraduate Sciences," pp. 70–106. See also W. H. G. Armytage, *Civic Universities. Aspects of a British Tradition* (London: Ernest Benn, 1955), chaps. 4, 5, and 6 on "The Emergence of a Scientific Spirit," "The Baconian Blueprint," and "The Royal Society and its Influences." A more general discussion of education in the period will be found in J. W. Adamson's *Pioneers of Modern Education* (Cambridge, England: Cambridge University Press, 1921). His study includes Bacon, Hartlib, Comenius, Dury, Milton, and Petty.

12. Abraham Cowley, *A Proposition for the Advancement of Experimental Philosophy*, in *Complete Works in Verse and Prose*, ed. Rev. Alexander B. Grosart (Edinburgh, 1881), II, preface, p. 285.

13. *Ibid.*

14. Richard F. Jones, *Ancients and Moderns, A Study of the Background of the Battle of the Books* (Saint Louis, Missouri: Washington University Studies, 1936), gives the history of this well-known quarrel. He shows that it was the scientific progress of the moderns that won the battle and dethroned the ancients by 1672, the date of Isaac Newton's first communication to the Royal Society.

15. J. B. Bury, *The Idea of Progress* (New York: Dover, 1955), is one of the general works in the background of the present study. See especially chap. 2, "Utility the End of Knowledge: Bacon," pp. 50–63, and chap. 4, pp. 78–97, for a recapitulation of "The Doctrine of Degeneration: The Ancients and Moderns."

16. *Divine Comedy*, trans. Laurence Binyon (New York: Viking Press, 1955), *Purgatorio*, 3.34.

17. In discussing Cowley's attitude toward this question Robert B. Hinman, *Abraham Cowley's World of Order* (Cambridge, Massachusetts: Harvard University Press, 1960), p. 141, carries the

quotation from Agrippa, *Of the Vanitie and Uncertaintie of Artes and Sciences*, trans. James Sanford (London, 1676), pp. 365–368.

18. Douglas Bush, "Two Roads to Truth: Science and Religion in the Early 17th Century," ELH, VIII (1941), 81–102.

19. On February 26, 1616, the Sacred Congregation of the Roman Catholic Church officially suspended *De Revolutionibus* until corrections were made. See Giorgio de Santillana, *The Crime of Galileo* (Chicago, Illinois: University of Chicago Press, 1955), for an excellent treatment of this and Galileo's position with the Catholic Church.

20. *The Defense of Galileo of Thomas Campanella*, trans., introd., and ed. Grant McColley in *Smith College Studies in History*, XXII (April–July 1937), introd., pp. 7–8.

21. *Anatomy of Melancholy* (London: Dent, 1949), II, "Digression of Air," pp. 55–56. I have included only the English translations of Burton's Latin phrases in this passage.

22. Ernest Lee Tuveson, *Millennium and Utopia: A Study in the Background of the Idea of Progress* (Berkeley and Los Angeles, California: University of California Press, 1949), preface, pp. 7–8.

23. For a basic and thorough interpretation of attitudes toward religion and science in this century, see E. A. Burtt, *The Metaphysical Foundations of Modern Physical Science* (New York: Doubleday, 1954), Richard S. Westfall, *Science and Religion in Seventeenth-Century England* (New Haven, Connecticut: Yale University Press, 1958), and Paul H. Kocher, *Science and Religion in Elizabethan England* (San Marino, California: Huntington Library, 1953). The last study concentrates on the period 1550–1610 and is especially interesting in the discussion of physicians' attitudes toward science and religion, chaps. 11–13, pp. 225–283.

24. George Sarton, *Six Wings: Men of Science in the Renaissance* (Bloomington, Indiana: Indiana University Press, 1957), p. 62.

25. *Christianopolis*, chap. 62, "Geometry," p. 221.

26. *Ibid.*, chap. 63, "Mystic Numbers," pp. 221–222.

27. *Ibid.*, chap. 36, "The Director of Learning," p. 187. (Italics mine.)

28. *Ibid.*, chap. 49, "Mathematical Instruments," p. 203. (Italics mine.)

29. *Ibid.*, chap. 70, "Natural Science," pp. 231–232.

30. Tommaso Campanella, *Città del Sole*, trans. Daniel J. Donno (see note 9 in Chapter IV), pp. 59–60. This unpublished translation

is subsequently referred to as *City*. There are two easily available discussions of Campanella's theories of revelation: Harald Höffding, *A History of Modern Philosophy*, trans. B. E. Meyer (New York: Humanities Press, 1950), I, 149–158, and Alfred Weber, *History of Philosophy*, trans. Frank Thilly (New York, 1896), pp. 291–295.

31. *City*, p. 62.

32. *Ibid.*

33. Mary Louise Berneri, *Journey Through Utopia* (Boston, Massachusetts: Beacon Press, 1951), p. 128.

34. "Of Unity in Religion," in *Francis Bacon*, ed. R. F. Jones (New York: Odyssey Press, 1937), p. 8.

35. *Novum Organum*, in *The Works of Francis Bacon*, ed. Spedding, Ellis, and Heath (London, 1887), IV, 1.65. (Italics mine.) Subsequent references to Bacon's works are to this edition.

36. Stephen Penrose, in the first chapter of his work, *The Reputation and Influence of Francis Bacon in the 17th Century* (diss., Columbia University, privately printed, 1934), discusses this. He explains Ralph Cudworth's opposition to Bacon on anthropomorphism and on his position regarding first causes. Penrose says that Cudworth never called Bacon an atheist; however, the charge was made, and perhaps earlier than is usually thought, because William Rawley felt the need to defend him from this criticism.

37. Benjamin Farrington, *Francis Bacon: Philosopher of Industrial Science* (New York: Henry Schuman, 1949), pp. 3–31.

38. *New Atlantis*, in *The Works of Francis Bacon*, III, 137.

39. Christopher Hill emphasizes the anticlerical attitude in *The Law of Freedom* . . . in a similar manner in his introduction to *Gerrard Winstanley: Selections from His Works*, ed. Leonard Hamilton (London: Cresset Press, 1944). Hill's interest, however, is in showing Winstanley as an English precursor of Marxist theory, which is not our concern in this study. For a more objective evaluation of Winstanley's attitudes, the position of religion in his writings as a whole, and his advocacy of common ownership of land, see W. Schenk, *The Concern for Social Justice in the Puritan Revolution* (London: Longmans Green, 1948), chap. 6, "Gerrard Winstanley, the Digger," pp. 97–113, and Winthrop S. Hudson, "Economic and Social Thought of Gerrard Winstanley. Was he a Seventeenth-Century Marxist?" in R. L. Schuyler and H. Ausubel, *The Making of English History* (New York: Dryden Press, 1952), pp. 303–318.

40. Winstanley, p. 565.

41. *Science and Religion,* pp. 132–134.

42. Joseph Glanvill, "Anti-fanatical Religion, and Free Philosophy. In a Continuation of the New Atlantis," *Essays on Several Important Subjects in Philosophy and Religion* (London, 1676), pp. 54–55. This essay is cited as Glanvill, "Continuation." See Richard Westfall, *Science and Religion,* pp. 46–47, 59, 115–118, 130–131, 174–183, for discussions of Glanvill's tenets in his other writings as well as the "Continuation."

43. Glanvill, "Continuation," pp. 44–45.

44. Basil Willey, *Seventeenth Century Background,* p. 174.

VII : THE RIGHT METHOD

1. G. H. Turnbull, *Hartlib, Dury and Comenius* (London: Hodder and Stoughton, 1947), pp. 36–37. A footnote explains that Reineri was "presumably Henry Reneri, a Professor of Philosophy at Deventer in November 1632 according to a letter from De la Greue to Dury." In the Hartlib papers there were also found two documents of Acontius, entitled *De Praefatione Acontii in Methodum and Succincta Delineatio rerum in Acontii methodo contentarum.*

2. It is difficult to say whether this was the work enclosed. The *Idea of Mathematics* is generally thought to have been written nine years later, about 1639. John Pell, however, had numerous manuscripts on the subject, and this may have been an early one. The *Idea* was especially interesting to Hartlib, who sent copies of it to Mersenne and Descartes. Their comments on it were later published with it by Robert Hooke in *Philosophical Collections* (1679).

3. A. R. Hall, *The Scientific Revolution 1500–1800: The Formation of the Modern Scientific Attitude* (Boston: Beacon Press, 1956), p. 160.

4. Alfred N. Whitehead, *Science and the Modern World* (New York: Mentor Books, 1956), chap. 2, "Mathematics as an Element in the History of Thought," pp. 20–39.

5. *Johann Valentin Andreae's Christianopolis,* trans. and introd. Felix E. Held (diss., University of Illinois, privately printed, 1914), chap. 61, "Arithmetic," p. 219. The work is cited as *Christianopolis.*

6. *Ibid.,* chap. 62, "Geometry," pp. 220–221.

7. *Ibid.,* chap. 49, "Mathematical Instruments," p. 204.

8. Charles Singer, *A Short History of Scientific Ideas* (Oxford: Clarendon Press, 1959), p. 42.

9. Benjamin Farrington, *Science in Antiquity* (London: Oxford University Press, 1950), p. 117.

10. Alfred Weber, *History of Philosophy*, trans. Frank Thilly (New York, 1896), p. 294.

11. Basil Willey, *The Seventeenth Century Background* (New York: Doubleday Anchor, 1953), p. 187, quotes this from *The Vanity of Dogmatizing* and shows its origin in Hobbes' *Leviathan*, chap. 4, in which Hobbes says that geometry is "the only science that it hath pleased God hitherto to bestow on mankind."

12. Loren Eiseley, *Francis Bacon and the Modern Dilemma* (Lincoln, Nebraska: University of Nebraska Press, 1962), p. 12. Eiseley cites this as evidence that Bacon wrote like a "modern theoretical physicist." Fulton H. Anderson, *The Philosophy of Francis Bacon* (Chicago, Illinois: University of Chicago Press, 1948), very carefully shows Bacon's attitude toward mathematics (pp. 163–164, 192–193, 281, 292, 301, 302) and as a handmaid of physics (pp. 121, 166).

13. J. W. Adamson, *Pioneers of Modern Education* (Cambridge, England: Cambridge University Press, 1921), pp. 9–10.

14. [R. H.], *New Atlantis. Begun by the Lord Verulam, Viscount St. Albans: and Continued by R. H. Esquire* . . . (London, 1660), pp. 65–66. Subsequent references to this will be cited as R. H. J. Neper or Napier (1550–1617) was a Scottish mathematician who first used the word "logarithm" for his publication in 1614. His other works included the *Rabdologiae* (1617), which contained various methods of abbreviating arithmetical processes, such as Napier's rods or bones based on multiples of numbers, and the *Constructio* (1619), printed posthumously by his son Robert and edited by Henry Briggs. In this Napier introduced the decimal point in writing numbers.

Johannes Regiomontanus (Johann Muller, 1436–1476) was a notable German astronomer who, with the wealthy Bernard Walther, is reputed to have equipped the first European observatory. He recorded observations of the comet (Halley's) of 1472 that supplied a basis for modern cometary astronomy. In mathematics he contributed a treatise, *De Triangulis* (1533), a landmark in the development of modern trigonometry. Among the various instruments he devised were the wooden eagle and the iron fly, which gained notoriety as flying automata. John Wilkins, *Mathematicall Magick* (London, 1648), II, 6.191–199, described their operation as guided boomerangs. Marjorie Nicolson quotes 17th-century poetic references to the eagle and the fly in her *Voyages to the Moon* (New York: Macmillan, 1948), pp. 151–152.

Erasmus Reinhold (1511–1553) presented the first table of tangents in his *Canon Foecundus,* printed in the year of his death. Earlier he had published the *Tabulae Prutenicae,* calculated on Copernican principles, and in his edition of G. Peurbach's *Theoricae Novae Planetarum* (1542) he described the optical apparatus of the camera obscura, which he used for solar observations. An instrument of this type was later used by Brahe, Mästlin, and Johann Kepler.

Thomas Harriot (Harriott) was a prominent English mathematician and astronomer (1560–1621). He assisted the development of algebra, introducing symbols and notation still used, for example, x^3 for xxx, $>$ for "greater than," and $<$ for "less than." His *Artis analyticae praxis ad sequationes algebraicas resolvendas* was published at London in 1631.

15. Professor A. R. Hall, in *The Scientific Revolution,* summarizes clearly the earlier use of the experimental method by Roger Bacon and William of Ockham's philosophic interest in it. But Hall shows that this sort of interest was very different from an application to scientific problems. Furthermore, he claims that "no important re-statement of scientific method was made during the sixteenth century." It was in the next era that the traditional ideas failed to satisfy the majority (p. 163).

16. Willey, *The Seventeenth Century,* p. 187. The quotation is from *The Vanity of Dogmatizing.*

17. From the Proem to the 1565 edition of Telesio's *De rerum natura,* quoted by Edmund G. Gardner, *Tommaso Campanella and His Poetry* (Oxford: Clarendon Press, 1923), p. 4.

18. Harald Höffding, *A History of Modern Philosophy,* trans. B. E. Meyer (New York: Humanities Press, 1950), I, 153–154. Cf. Weber, *History,* pp. 291ff.

19. Abraham Cowley, *A Proposition for the Advancement of Experimental Philosophy,* in *Complete Works in Verse and Prose,* ed. Rev. Alexander B. Grosart (Edinburgh, 1881), II, preface, p. 286. The work is cited as Cowley, *Proposition.*

20. *Novum Organum,* in *The Works of Francis Bacon,* ed. Spedding, Ellis, and Heath (London, 1887), IV, 1.51. Subsequent references to Bacon's works will be to this edition.

21. *John Amos Comenius: Selections,* introd. Jean Piaget (Switzerland: Centrale Lausanne S.A., 1957), pp. 79–80.

22. Robert Hooke, "The Present State of Natural Philosophy, and wherein it is deficient," *The Posthumous Works of Robert Hooke* (London, 1705), p. 6.

23. *Science and the Modern World*, pp. 42–43. Cf. Loren Eiseley, *Francis Bacon*, and René Dubos, *The Dreams of Reason: Science and Utopia* (New York: Columbia University Press, 1961). Dubos presents an excellent philosophic evaluation of man's expectations written from a scientific point of view. Bacon's *New Atlantis* is the example for utopias.

24. Singer, *A Short History*, pp. 264–267.

25. Eiseley, *Francis Bacon*, also mentions this (p. 31) with regard to Lord Acton's opinion.

26. *Novum Organum*, IV, 1.61.

27. *New Atlantis*, in *The Works of Francis Bacon*, III, 164. Here Bacon does not go on to include aids to the intellect and the furthering of practical operation as he did in the *Novum Organum*. He seems to assume these in the fictional utopia. See Fulton H. Anderson, *The Philosophy of Francis Bacon*, chaps. 20 and 21, pp. 229–258.

28. *Christianopolis*, chap. 11, "Metals and Minerals," p. 154.

29. *Ibid.*

30. William Davenant, *Gondibert*, in *The Works of Sir William Davenant* (London, 1672), bk. 2.5.14.

31. Cowley, *Proposition*, p. 288.

32. A. R. Hall, *The Scientific Revolution*, p. 167.

33. Farrington, *Science in Antiquity*, p. 175.

34. See Singer, *A Short History*, pp. 207–210, for details concerning these men.

35. *Christianopolis*, chap. 45, "The Drug Supply House," p. 198.

36. *Ibid.*, chap. 47, "The Natural Science Laboratory," p. 200.

37. Other writers in the early part of the century were also accepting Tycho's proof. Robert Burton, *The Anatomy of Melancholy* (London: Dent, 1949), II, 48, mentioned Cardan, Tycho, and John Pena's proof by refractions that there was no element of fire. And, of course, John Donne's famous line in the *First Anniversary* comes to mind: "The Element of fire is quite put out."

38. Tommaso Campanella, *Città del Sole*, trans. Daniel J. Donno (see note 9 in Chapter IV), p. 6. This unpublished translation is subsequently referred to as *City*.

39. Plato, *Laws*, 12.950, 952. In these sections on foreign relations, the ambassador is to be no less than 40 years of age and he is to be

selected as a representative to make a good impression at the games and sacrifices of the nation visited. Other spectators sent in the public's behalf will observe the customs and perhaps know those "few inspired men" in the world from whom they will learn much. The ambassador reports to an august assembly at home, and if they decide that some foreign custom or knowledge is superior to their own, they may transmit it to their society and "the younger men shall learn" it.

40. This quotation is from the first edition of 1621 (p. 58), but Burton does not cite the source in a marginal note until the 1624 edition, the second of his many editions. Here he says, "Plato, 12 de legibus, 40 annos natos vult. ut si quid memorabile viderint apud exteros, hoc ipsum in rempub. recipiatur."

41. Cowley, *Proposition,* p. 288.

42. Winstanley was slightly ahead of actual developments in the postal service of England. The government had sporadically instituted controls for foreign mail, largely to prevent dangerous schemes from being instigated abroad. Various reforms were attempted in the reign of Charles I to improve inland services, but it was not until Oliver Cromwell's Post Office Act of 1657 that a comprehensive system with a postmaster general was established.

43. Gerrard Winstanley, *The Law of Freedom in a Platform, or True Magistracy Restored,* in *The Works of Gerrard Winstanley,* ed. and introd. George H. Sabine (Ithaca, New York: Cornell University Press, 1941), p. 571.

44. W. H. Armytage, *Civic Universities: Aspects of a British Tradition* (London: Ernest Benn, 1955), p. 102.

45. Samuel Hartlib, "A Further Discovery of the Office of Publick Address for Accommodations," *Harleian Miscellany* (London, 1810), VI, 14–27. First printed as a quarto containing 34 pages in 1648.

46. William Petty, *The Advice of W. P. to Mr. Samuel Hartlib, for the Advancement of some particular Parts of Learning,* in *Harleian Miscellany* (London, 1810), VI, 2.

47. "Vellus Aureum: sive Facultatum Luciferarum Descriptio magna." The translation is my own.

48. See A. R. J. P. Ubbelohde's essay in full, "The Beginnings of the Change from Craft Mystery to Science as a Basis for Technology," *A History of Technology,* ed. Charles Singer *et al.* (Oxford: Clarendon Press, 1958), IV, 663–681.

VIII : INVENTIONS AND RESULTS

1. Trans. Henry Crew and Alfonso De Salvio, introd. Antonio Favaro (Evanston, Illinois: Northwestern University Press, 1950), p. 1. The original text (Leyden, 1638) is here translated.

2. Trans. Desmond I. Vesey (London: Methuen, 1963), scene 1, p. 30.

3. Bacon makes this point repeatedly. Cf. Benjamin Farrington, *Francis Bacon: Philosopher of Industrial Science* (New York: Henry Schuman, 1949), p. 45.

4. *Novum Organum*, in *The Works of Francis Bacon*, ed. Spedding, Ellis, and Heath (London, 1887), IV, 1.129. Subsequent references to Bacon's works will be to this edition.

5. Joseph Glanvill, *Essays on Several Important Subjects in Philosophy and Religion* (London, 1676), Essay III, pp. 1–56.

6. Galileo and Descartes had also disputed the doctrine of a physical vacuum, and there were, of course, many men who had contributed to the development of the air pump. The instrument is usually attributed to Otto von Guericke in 1650. Robert Boyle and Robert Hooke both worked to perfect it.

7. *New Atlantis*, in *The Works of Francis Bacon*, III, 162.

8. Marjorie H. Nicolson, "The Microscope and English Imagination," *Science and Imagination* (Ithaca, New York: Cornell University Press, 1956), p. 158. The reader is referred to this article for a full description of the instrument, its origins, and its effects on the literature of the period.

9. Selenology is not to be confused with selenography, which, in contradistinction, is the description of the moon's surface.

10. Professor Daniel Donno, at my request, has researched the question and concludes with Firpo and Bobbio that this reference was added to the original text after 1611 when Campanella had seen Galileo's *Nuncius*. Firpo has reviewed the genealogy of MSS. (*Giornale Storico della letteratura Italiana*, CXXV, 1948, pp. 225ff) and finds the reference to the "glass" in the 1611 (Trent) Italian MS. and in two subsequent Latin MSS. (1623 and 1637). It is not in the original 1602 MS. reprinted by Solmi in 1904.

11. It is curious to note that in Bacon's experimental houses each of the five senses has its special research quarters except the tactile sense, which he omits.

12. The work has been republished by Grant McColley in *Smith College Studies in Modern Languages*, XIX (October 1937), 3–78. It is an article describing man's efforts to communicate by birds, smoke signals, horns, and finally Godwin's new device.

13. [R. H.], *New Atlantis. Begun by the Lord Verulam, Viscount St. Albans: and Continued by R. H. Esquire* (London, 1660), pp. 67–68. Subsequent references will be cited as R. H.

14. Robert K. Merton, in "Science, Technology, and Society in Seventeenth-Century England," *Osiris*, IV (1938), 360–362, examines sociologic factors as an important influence if not a determinant of man's areas of interest and activity. Regardless of the issue of determinism, Merton's study reveals emphasis on these same categories.

15. *Ibid.*, p. 571.

16. Great Britain, Office of the Commissioners of Patents for Inventions, *Patents for Inventions. Abridgments of Specifications Relating to Agriculture. Division I. Field Implements (Including Method of Tilling and Irrigating Land) 1618–1866* (London, 1876), III, 1. At the end of the entry from which the quotation is taken is the statement, "No specifications enrolled." Not until the mid-18th century are specifications given in more detail in these records.

17. *Ibid.*, p. 3. A list of the interests and activities of Hartlib's friends employs the same generalized statement and frequently claims remarkable results. It reads like the utopists' writings. See G. H. Turnbull, "Samuel Hartlib's Influence on the Early History of the Royal Society," *Notes and Records of the Royal Society*, X (1953), 112–118.

18. R. H., p. 65. I am unable to identify Boniger.

19. Trans. Gilstrap in Glenn Negley and J. Max Patrick, *The Quest for Utopia* (New York: Henry Schuman, 1952), p. 335. This translation is based on the 1623 Latin edition.

20. *New Atlantis*, pp. 158–159.

21. Sir Hugh Plat, *Delightes for Ladies* (London, 1609), reprinted with introd. G. E. Fussell and Kathleen Fussell (London: Crosby Lockwood, 1948).

22. R. H., p. 80. Marle (also spelled marl) is a kind of soil composed of clay with carbonate of lime and known for use as a fertilizer. Milton mentions the "burning marl" in his description of hell's torments in *Paradise Lost*, 1.296. Brakes are fern or bracken.

23. Robert K. Merton, "Science, Technology, and Society," pp.

536–538, describes the seriousness of this problem in England at the time.

24. Thomas Sprat, *History of the Royal Society*, ed. Jackson I. Cope and Harold W. Jones (Saint Louis, Missouri: Washington University Studies, 1959), pp. 149–150.

25. Thomas Birch, *The History of the Royal Society of London for improving of Natural Knowledge* . . . (London, 1756), I, 460. Other men involved in this were Goddard, Merret, Winthrop, Ent, and Willughby.

26. The 1623 edition of the *City of the Sun* has greatly enlarged passages on illness, foods with medicinal values, and cures somewhat altered from earlier MSS.

27. *New Atlantis*, pp. 160–161.

28. *Johann Valentin Andreae's Christianopolis*, trans. and introd. Felix E. Held (diss., University of Illinois, privately printed, 1914), chap. 98, "The Sick," p. 274. The work is cited as *Christianopolis*. Robert Burton's modern approach to psychiatry has been lauded by William Osler and Bergen Evans, but their evaluations were based on *The Anatomy of Melancholy* as a whole, not on the utopia in the preface, which does not specifically treat this matter.

29. Benjamin Farrington, *Science in Antiquity* (London: Oxford University Press, 1950), p. 93.

30. R. H., p. 53. There is an error in pagination in this edition, resulting in two pages of this same number, but the text is consecutive.

31. Merton, "Science, Technology, and Society," p. 383, cites as his main authority for these statements Johann H. Baas, *Outlines of the History of Medicine and the Medical Profession*, trans. H. E. Henderson (New York, 1889).

32. Lewis Mumford, *The Story of Utopias* (New York: P. Smith, 1941), suggests the former view and Mary Louise Berneri, *Journey Through Utopia* (Boston, Massachusetts: Beacon Press, 1951), holds the latter position.

33. *Christianopolis*, chap. 11. "Metals and Minerals," p. 154.

34. Gerrard Winstanley, *The Law of Freedom in a Platform* . . . , in *The Works of Gerrard Winstanley*, ed. and introd. George H. Sabine (Ithaca, New York: Cornell University Press, 1941), pp. 579–580.

35. Merton, "Science, Technology, and Society," p. 502, states further that examination of patent statistics for the period 1561–1688

shows that about 75 percent of 317 patents were directly or indirectly related to this industry.

36. R. H., pp. 68–71. R. H. may well have used John Wilkins' *Mathematicall Magick or, The Wonders that may be performed by Mechanical Geometry* (London, 1648), as a source for many of the inventions he described. Wilkins mentions each of these, with the exception of petrified ice, as follows: 1. the malleable glass made by an artificer under the reign of Tiberius (bk. 1, chap. 1, p. 6); 2. an experiment in *aurum fulminans*, "one scruple of which shall give a lowder blow, & be of greater force in descent, then half a pound of ordinary gunpowder in ascent" (bk. 2, chap. 12, p. 255); 3. a lamp burning constantly in a hollow vessel stopped up at all vents and possible chemical compositions to produce this effect (bk. 2, chap. 12, pp. 250–251). Wilkins includes a description of Archimedes' sphere which R. H. also describes. See p. 255 in this study.

37. R. H. was exceedingly fond of vitriol in his various formulas. Vitriol is one of various native or artificial sulfates of metals used in the arts or medicine, especially sulfate of iron. Oil of vitriol was a concentrated sulfuric acid, and spirit of vitriol is the distilled essence (*OED*). To make a "Sympatheticall powder," which he noted was in use in Europe, R. H. called for only the purest "vitriol calcined white in the sun" and the gum of tragacanth. This was a "most salubrious balsom" for wounds; and, like the famous sympathetic powder of Sir Kenelm Digby, it could cure a wound many miles away from it.

38. Wilkins, *Mathematicall Magick*, bk. 2, chap. 12, pp. 246–247. Wilkins cites Plutarch and Pliny as his sources, and also mentions Bacon's interest.

39. *Sylva Sylvarum: or A Natural History*, in *The Works of Francis Bacon*, II, 591. This is the "Experiment solitary touching fuel that consumeth little or nothing."

40. Jackson Cope makes this observation in his introduction to Sprat's *History*, p. 21. The "oculus mundi" (eye of the world or the heavens) is a medieval term used for the variety of opal called Hydrophane (*OED*).

41. Sprat, *History*, p. 156.

42. R. H., p. 65.

43. Tommaso Campanella, *Città del Sole*, trans. Daniel J. Donno (see note 9 in Chapter IV), p. 32.

44. Merton, "Science, Technology, and Society," p. 539. Sprat's *History* also includes such projects.

45. Wilkins, *Mathematicall Magick*, bk. 2, chap. 5, pp. 178–190.

46. R. H., p. 82.

47. Simon Stevin (1548–1620) was notable as an early exponent of decimal arithmetic and for his experiments with DeGroot on the fall of heavy bodies to test Aristotle's theory of motion. He laid the "essential foundations for the whole science of hydrostatics" and analyzed into their simplest mathematical elements the traditional devices in the various trades, as for example, the cranes widely used in his native Netherlands. See Charles Singer, *A Short History of Scientific Ideas* (Oxford: Clarendon Press, 1959), pp. 223–224. Stevin is probably best remembered for his formal proof of the impossibility of perpetual motion. Stevin worked directly from Archimedes' theories, and one of his projects dealt with the laws of hydrostatic pressure in their application to the containing vessel. See George Sarton, *Six Wings: Men of Science in the Renaissance* (Bloomington, Indiana: Indiana University Press, 1957), p. 80. We may conclude that R. H. was well informed in selecting Stevin's invention for high honor.

48. Marjorie H. Nicolson, *Voyages to the Moon* (New York: Macmillan, 1948), pp. 150–151. The facts and inner quotation in this passage are from Wilkins, *Mathematicall Magick*, bk. 2, chap. 2, pp. 155–156. Wilkins states that, according to several credible authors, such chariots were commonly used on the "Champion plains of *China*" and also tried in Spain, though he does not know, in this instance, how successful the experiments were.

49. The poem "Inter Currus Veliferi" was published in *Hugonis Grotii Poemata Collecta olim a fratre ejus, Guil. Grotius* (London, 1639), pp. 385–396.

50. Wilkins, *Mathematicall Magick*, bk. 2, chap. 2, p. 154.

51. Cressy Dymock's tract, *An Invention of Engines of Motion lately brought to perfection. Whereby may be dispatched any work now done in England or elsewhere (especially Works that require strength and swiftness) either by Wind, Water, Cattel or Man* . . . (1651), has been thought incorrectly to be an anticipation of the steam engine according to H. Dircks, who reprints this quarto with *A Biographical Memoir of Samuel Hartlib* (London, 1865). Dymock, usually a writer on agricultural topics, and a corre-

spondent of Hartlib, claimed that he had a large-scale model of this engine at Lambeth. The tract is the usual evasive description with great claims for efficiency. It is hard to imagine anyone thinking of it in terms of the steam engine. Hartlib seems to have changed his mind about the invention. In a letter to Boyle (1654) he says he is "no more so fond as I was wont to be" about Dymock's idea. A similar claim was made by Edward Somerset, second Marquis of Worcester, in *A Century of the Names and Scantlings of such Inventions* . . . (London, 1663), for a method of raising water by means of steam. Again details are obscure and there is no illustration of the device. It was only a short time, however, until Thomas Newcomen's steam engine (1705) was in actual use for pumping water out of mines. Henry Beighton added automation to the process in 1718. This perfected device was then commonly used until James Watt's improvements were made, starting in 1763. Before Newcomen's invention, of course, were the basic contributions of Savery, Papin, Boyle, Huygens, and Galileo in the 17th century.

52. *New Atlantis*, pp. 157, 163–164.

53. Merton, "Science, Technology, and Society," pp. 522–534.

54. William Davenant, *Gondibert*, in *The Works of Sir William Davenant* (London, 1672), bk. 2, canto 6.29–40.

IX : UTOPIAS IN PERSPECTIVE

1. George Orwell, *Nineteen Eighty-Four* (New York: New American Library, 1956), p. 203. See also Chad Walsh, *From Utopia to Nightmare* (New York: Harper and Row, 1962).

2. Aldous Huxley, *Brave New World* (New York: Bantam Books, 1962), foreword, pp. ix–x.

3. Huxley carefully selected the name of his rejuvenating pill. Theodor H. Gaster, in *The Oldest Stories in the World* (Toronto, Canada: Viking, 1952), p. 51, says, "The soma of Indic belief is a kind of elixir of life contained in the juice of a paradisal plant." According to the *OED*, the soma plant yields a juice that is made into an intoxicating drink prominent in Vedic ritual.

4. Loren Eiseley, *Francis Bacon and the Modern Dilemma* (Lincoln, Nebraska: University of Nebraska Press, 1962), pp. 53–54, carries the quotation in similar context.

5. *Novum Organum*, in *The Works of Francis Bacon*, ed. Sped-

ding, Ellis, and Heath (London, 1887), IV, 1.129. (Italics mine.) Subsequent references to Bacon's works will be to this edition.

6. Tommaso Campanella, *Città del Sole*, trans. Daniel J. Donno (see note 9 in Chapter IV), p. 12. This unpublished translation is referred to as *City*.

7. Benjamin Farrington, *Francis Bacon: Philosopher of Industrial Science* (New York: Henry Schuman, 1949), p. 54.

8. *Novum Organum*, 1.129.

9. *New Atlantis*, in *The Works of Francis Bacon*, III, 165–166.

10. Samuel Hartlib, *A Description of the Famous Kingdom of Macaria*, in *Harleian Miscellany* (London, 1808), I, 582.

11. Abraham Cowley, *A Proposition for the Advancement of Experimental Philosophy*, in *Complete Works in Verse and Prose*, ed. Rev. Alexander B. Grosart (Edinburgh, 1881), II, 288.

12. *Ibid.*, p. 289. While he rewarded generously, Cowley also was concerned about supporting the college, so he stipulated that, if the invention was profitable on the market, one-third part of the return should belong to the inventor and the "two other to the Society."

13. *Novum Organum*, 1.129.

14. On this point R. H., in the *New Atlantis. Begun by the Lord Verulam . . .* (London, 1660), p. 53, differed sharply from Bacon's secretive attitude regarding inventions. Bacon's fellows in Salomon's House revealed only what they wished or thought best for the state. In R. H.'s account, no monopoly or closeting of inventions was permitted. "We study the publick good so much, that whereas we reward those that discover, so he is in some measure punished that conceals and hides a benefit which may pleasure his countrey."

15. William L. Laurence, "Science in Review," *New York Times* (January 11, 1959), sec. 4, p. 11. Mr. Laurence explains that the principle of Lunik's orbit was discovered by Newton.

16. *The Works of H. G. Wells* (New York: Charles Scribner's Sons, 1925), IX, 245.

17. *Ibid.*, p. 92. See also Gerard Piel, *Science in the Cause of Man* (New York: Random House, 1964), chap. 5, "The New Paradise," pp. 134–151, and Don K. Price, *The Scientific Estate* (Cambridge, Massachusetts: Harvard University Press, 1965), chap. 1, "Escape to the Endless Frontier," pp. 1–20.

18. Bertrand Russell, *Icarus or the Future of Science* (New York: E. P. Dutton, 1924), pp. 62–63.

19. Dennis Gabor, "Inventing the Future," *Encounter* (May 1960), p. 15. The quotation following is from his book of the same name (New York: Knopf, 1964), p. 207. The main quotation was carried by Richard Schlatter in his article, "Humanistic Scholarship in the United States," *American Council of Learned Societies' Newsletter*, XV (January 1964), 4–5.

20. Oscar Wilde, "The Soul of Man under Socialism," in *The Works of Oscar Wilde*, introd. Richard Le Gallienne (New York: Lamb Publishing Co., 1909), VIII, 148.

INDEX